1.0 RESEARCH OBJECTIVES

The goal of the Advanced Concepts effort at the Air Force Research Lab was to "enable future Air Force missions through the discovery and demonstration of emerging revolutionary technology." The research was primarily focused on propulsion and was divided by application into concepts for space-access, near-space, and in-space. The Advanced Concepts effort was a technology push effort and was not limited by strict ties to specific needs of near-term programs. Research was focused on technologies that could potentially be fielded in the 15-50 year time frame. Research was also limited to topics that were based on proven physics.

The Advanced Concepts effort used a general process to address specific identified problems in each of the application areas: space access, near-space, and in-space. The first step was to identify and quantify specific key challenges in the application areas that, if solved, would lead to a revolutionary advancement in mission capabilities. Thorough reviews of the application area and potential advanced concepts within the area were then conducted (if they were not already available). The reviews yielded: a general understanding of relevant previous work, typical hardships encountered in similar work, envisioned near-term improvements, and candidate advanced concepts technologies. The final step was to conduct targeted research using both experimental and computational methods to address key unknowns with the potential advanced concepts to determine if they could meet the requirements for solving the challenges. Generally an attempt was made to advance the technologies to a technology readiness level (TRL) of approximately three (basic proof-of-concept).

This report summarizes the efforts taken in each of the application areas: space access, near-space, and in-space propulsion. An overview of the application area is given along with the key technological challenges identified for the application area. Summaries of the conducted reviews are given and the associated publications for the reviews are referenced for more information. Brief summaries of the targeted research projects on individual concepts are also given. The current status and technological breakthroughs needed for each concept to be able to achieve the required performance will also be given for future reference. One or more of the key publications produced during the research on each individual concept are also referenced for more information.

2.0 ADVANCED CONCEPTS FOR SPACE ACCESS

The means of ferrying every man-made object taken from the ground to space has been through chemical combustion of one type or another. From the days of Sputnik, launched with a combination of liquid oxygen and kerosene, to modern multistage launch vehicles, the paradigm has been the same: combine fuel and oxidizer to extract chemical energy that is then converted to kinetic energy. Current chemical propulsion technology is very efficient, on the order of 97–98%. For example, the space shuttle main engines (SSME) are approximately 97% efficient, leaving little room for additional performance improvements. Steady, incremental improvements in propellants and structural materials can yield evolutionary improvements in performance, but will not yield the desired revolutionary improvements.

The first step in the Advanced Concepts for Space Access effort was to identify the critical space access metric. There are many cost and performance metrics that are commonly used to characterize the launch process. Which one, if significantly improved, would yield a revolutionary effect on the space access process? Current launch systems have delivered 120,000 kg to LEO (Saturn V, flown 1967 – 1972), have sent a payload beyond our solar system (Voyager I, launched in 1977 and at 120 AU as of Aug, 2012), have sustained a high reliability (R > 0.95)[1], and some (particularly solid rocket motor based systems) are responsive and could be launched with several days of notice. In general, none of these factors has significantly improved in more than three decades and improving them, although useful, would not have a revolutionary effect on the launch process. It is widely recognized that the single most important metric for space access is the cost of launch. NASA identified the top Technological Challenge for all technology objectives were: "improved access to space" with "dramatically reduced total cost."[2] Similar to the other metrics, launch costs have been relatively stable with few signs indicating that rocket based launch systems can achieve the required reductions in cost at the current flight rates.

A widely recognized goal for space access is to "reduce launch costs by an order of magnitude for the entire launch spectrum". It is very important, however, to add "at the current launch rate" to that statement. Launch rate has a very significant effect on the launch costs and its effects must be included when comparing different launch concepts.[3] Concepts that promise reductions in launch costs over the existing state-of-the-art, but that achieve them by operating with daily launches verses the current rate of several per year per vehicle are not using a fair comparison. It is feasible that launch costs could be reduced by an order of magnitude simply by significantly increasing the launch rate.

There are good reasons to set the desired reduction in launch cost at an order of magnitude. National Aeronautics and Space Administration (NASA) market analysis has shown that the launch market is mostly inelastic until launch costs are lowered to roughly 1/10 of the current values.[4] Once launch costs are reduced to approximately 1/10 of their current value, the market will expand and more launches will be purchased which will tend to further reduce the launch costs. Several proposed future applications of space such as space based solar power[5], space tourism[6], and space mining[7] require similar improvements.

The Advanced Concepts effort identified one key challenge for space access.

- *Reduce the cost of access to space for the entire launch spectrum by one order of magnitude at current launch rates.*

The Advanced Concepts effort addressed this problem by first performing a thorough review of advanced concepts for space access. Several promising long term candidate technologies were identified including: gun launch, microwave and laser beamed energy propulsion (BEP), and high performance upper stages such as solar thermal propulsion systems. Only one of the BEP concepts, microwave thrust augmentation of solid rocket motors, was pursued due to budgetary and time constraints. Microwave thrust augmentation of solid rocket motors was evaluated because it was a unique, unexplored concept and represented perhaps the nearest term beamed energy propulsion concept. The concept appeared technologically viable, but suffered from the same limitation as all beamed energy launch systems: a very costly ground station.

2.1 Review of Advanced Concepts for Space Access[8]

The means of ferrying every man-made object taken from the ground to space has been through chemical combustion of one type or another. There is little room for further technological improvements to the current range of rocket based launch systems and the improvements that are made generally come at great expense. Current launch vehicles, however, place only a small percentage of their total lift-off weight into orbit, convert a small fraction of the stored chemical energy into payload mechanical energy, and cost significantly more than simply the energy and materials required for launch. Using a typical ΔV of 9.5 km/s as an example, the energy required to reach a low-Earth orbit is approximately 45 MJ/kg or 12.5 kW-hr/kg. This equates to \$1.75/kg at current peak-hour electricity rates and about \$0.48/kg at off-hour rates. Current rates for access to space range from several thousand to well over 10,000 US dollars per kilogram.[8] Clearly, there is also room for improvement in the area of cost for current chemical launch vehicles.

For decades, advanced propulsion concepts have sought alternative means to more easily and cost effectively access space; however, all of these concepts have yet to be effectively realized. The review evaluated a wide range of advanced concepts for space access for their potential for reducing launch costs for both launch and nanolaunch (payloads under 100 kg) within the next 15 to 50 years. The advanced concepts were compared to two existing systems that can fulfill these roles: the Delta IV Heavy[9] from the United Launch Alliance and the Minotaur IV[10] from Orbital Sciences Corporation.

Before evaluating the concepts for their potential to reduce launch costs it was first important to gain some understanding of what actually leads to such high launch costs. There have been many significant efforts to determine the causes of the high cost of access to space.[11,12] A simplified analysis performed by Taylor concluded that the three important causes are the amortized cost of research and development, the cost of constructing the vehicle itself, and the cost of the ground operations for the launch process.[13] It has also been shown that the launch rate also heavily affects the overall cost.[3] One thing that typically does not affect the launch cost is the cost of the propellant itself because it usually represents a very small fraction of the overall launch costs.

Figure 1 shows the distribution of current launch costs for common US launch systems (as of 2007) as a function of mass for the case of launching a satellite into a 185 km (100 nmi) circular orbit with an inclination of 28.5 degrees.[8] In general the cost per unit mass decreases as the delivered payload mass increases, but achieved launch system costs (at current launch rates) are between \$9,000/kg and \$50,000/kg (\$4,000/lb and \$23,000/lb). Launching payloads with masses up to approximately 100 kg on dedicated launchers have similar total cost since most of the cost is due to the launch process at these scales. This causes the cost/mass to significantly increase for very small payloads.

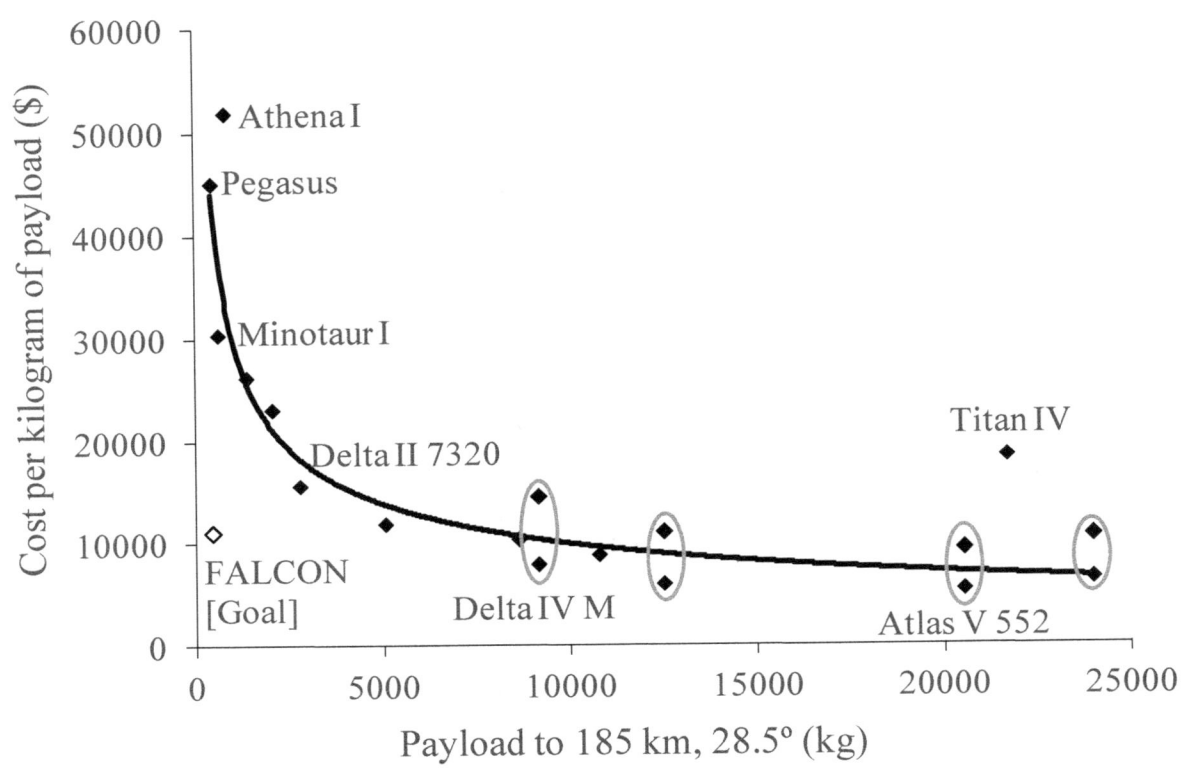

Figure 1. Launch Costs of Common Space Launch Systems[8]

In theory, launch vehicles using high energy density material (HEDM) propellants are the advanced launch concepts that are most similar to existing systems and would require minimal changes other than operating on a higher performance propellant. In practice, however, many platforms would require complete modifications to operate on some of the very high performance propellants that have been proposed. Reviews of the candidate HEDM propellants have been given elsewhere.[14] In general, it is advantageous for propellants to have high specific impulses for efficient mass utilization, but it is also important for them to have a high density for efficient storage. Clark has chronicled propellant development efforts in the United States.[15] The specific impulse is proportional to the square root of chamber temperature divided by the molecular weight of the exhaust, so if the molecular weight of the exhaust increases due to an increase in propellant density, chamber temperature must increase accordingly to maintain a given specific impulse. The relationship between specific impulse, temperature, and exhaust species molecular weight is shown in Figure 2.

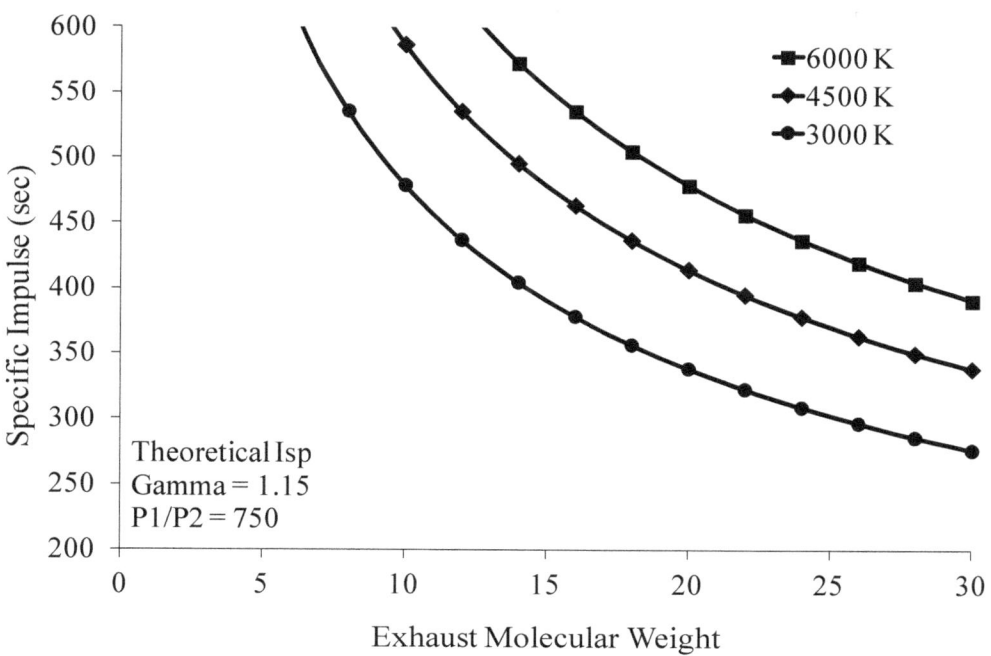

Figure 2. Specific Impulse as a Function of Temperature and Exhaust Molecular Weight

Advanced propellants are an attractive concept because their launch systems are similar to existing systems, but they must also meet a wide range of extra considerations including toxicity. Tripropellant combinations have demonstrated specific impulses above 500s which is sufficient for single stage to orbit (SSTO), but their exhaust was toxic which has stopped their use. A wide variety of propellants are being evaluated including strained ring hydrocarbons, polynitrogen, and even truly exotic theoretical propellants such as metallic hydrogen. Very high performance propellants will require significant developments in materials and methods to be viable. They are also typically less stable and likely to be expensive. None of the proposed propellants showed obvious ways that they would reduce costs. Estimates of costs for proposed systems indicate that they are at least as expensive as current systems (if proven technologically viable).

The next step further from existing launch vehicles is to vehicles that store the energy separately from the ejecta such as in nuclear thermal fission rockets. The state-of-the-art in nuclear thermal fission rockets were those developed during the ROVER/NERVA programs in the 1950s through the early 1970s. Nuclear thermal rockets were actually developed beyond the TRL of an advanced concept have been proven technologically viable. They have significantly lower thrust-to-weight ratios than chemical rockets and are only currently viable for a large upper stage engine. The high cost of their development and ground testing and operations makes it unlikely that they would achieve significant cost savings over existing systems for routine space access missions. Sociopolitical concerns also provide a hindrance to their development. Nuclear fission based space tugs that transfer spacecraft between orbits have also been evaluated and have similar limitations.

Another step away from the existing class of launch vehicles is achieved by removing the energy source completely from the vehicle and beaming the energy (either laser or microwave) from the ground to heat a propellant and produce thrust. Laser propulsion concepts can be broken down into four main categories based on the thrust mechanism employed: heat exchange, plasma formation (gas breakdown), laser ablation, and photon pressure. Microwave propulsion concepts can be divided into three categories that include heat exchanger or propellant heating, plasma formation, and coupling to solid rocket motor expansion gases/particles. The overriding advantage to either microwave or laser propulsion concepts is that the power source, and thus a large mass, is not integrated on the launch vehicle, remaining on the ground. The second advantage is the possibility for higher specific impulse since the limitation on chemical energy production is removed. The primary disadvantage with BEP concepts is that both the required laser and/or microwave sources (at the minimum practical power level) have just become available (as of 2012) and a ground station based on them is typically complex (usually needing 1000s of individual sources to produce the required ~1 GW) and typically very expensive (~$ billions). Several of the BEP concepts, particularly the heat exchanger variants, looked technologically viable, but none of them would achieve the required cost reductions without significant improvements in sources and alternative uses for the ground stations. One concept in particular, the mm-wave thermal rocket, looked appealing. Four technical challenges were identified for the concept and may be addressed with future work: heat exchanger, beam combining, tracking, and beaming high power density microwaves through the atmosphere.

Many other launch systems were also considered and are discussed more thoroughly in the review paper.[8] One system, gun launch, appeared technologically viable, and even financially advantageous, but appeared only able to launch gee tolerant payloads due to its high acceleration (~5000 gees). Other similar systems such as rail gun launch and coil gun launch may allow lower accelerations, but would require very long tracks and their energy per shot would need to be scaled by many orders of magnitude to be viable (only 33MJ as of 2010). Any concept that delivers all or most of the required ∆V at the Earth's surface will also suffer from extreme aerothermal loads.

Other more exotic concepts were also considered in the review. Space elevators are one concept that is regularly discussed. In a space elevator, a long tether is run between a massive counterweight, placed beyond Geosynchronous Earth Orbit (GEO), and the Earth's surface; a small "climber" is used to shuttle the payload up to orbit without using any propellant. Power is beamed to the climber using a laser. The appealing nature of the concept is that it does not require propellant or a disposable rocket at all. Some of the major concerns are: it is primarily applicable just to GEO, it requires a large mass to be placed beyond GEO, it requires a very long tether, it requires beamed power for days during "launch," and the tether has extreme material requirements. There are also concerns about collisions with micrometeoroids. Space platforms/towers where the launch occurs from the top of a very tall tower (~100 km) were also evaluated, but they appeared technologically problematic and appeared to provide very little performance benefit. Breakthrough physics based concepts were also evaluated, but none appeared technologically viable in the required timeframe or had any reason to expect significant reduction in costs.

The review identified no obvious candidate technologies for achieving the goal of *"reducing the cost of access to space for the entire launch spectrum by one order of magnitude at current launch rates"*. Gun launch is certainly technologically viable and could significantly reduce launch costs, but only for robust payloads such as propellant delivery that were able to withstand the approximately 5000 gee acceleration. Laser thermal and microwave thermal beamed energy systems appeared technologically viable, but would not reduce launch costs with current systems. Such systems must find alternative uses for the ground station and answer some fundamental technological questions before cost savings could be realized. High performance upper stages could also reduce launch costs in the relatively near future by allowing smaller first stage boosters to be used to launch a certain payload. They would not, however, achieve the goal of an order of magnitude reduction. A summary of the reviewed concepts for both launch and nanolaunch are given in Table 1 and Table 2, respectively. A color coded system is used to illustrate the technological feasibility (LTF), required magnitude of performance scaling (LMS), and current cost advantages (LCA) for the concepts.

Table 1. Summary of Advanced Concepts for Launch

Concept	LTF	LMS	LCA	Primary Challenge for Launch	Alternative Mission
Advanced Propellants				Many Diverse Requirements.	---
Air Breathing				Scramjet: Thermal Loads, Time-Scales.	---
Nuclear Thermal				Hydrogen Storage, Hot Hydrogen.	Space Tug
Solar Thermal				Hydrogen Storage, Hot Hydrogen.	Space Tug
Laser				Power Coupling, Ground Station Cost.	---
Microwave				Power Coupling, Ground Station Cost.	---
Gun Launch				High Gee Payloads, Aerothermal Loads.	Rapid, Robust Payload
Railgun				High Gee Payloads, Aerothermal Loads, Rails.	---
Space Platforms				Unfeasible.	---
Space Elevator				Materials, O, μmeteoroids, weather, vibrations...	Asteroid Mining
Breakthrough Physics				No known feasible concepts.	---

Table 2. Summary of Advanced Concepts for nanoLaunch

Concept	NTF	NMS	NCA	Primary Challenges for Launch	Alternative Mission
Advanced Propellants				Many Diverse Requirements.	---
Air Breathing				Scramjet: Thermal Loads, Time-Scales.	---
Nuclear Thermal				Hydrogen Storage, Hot Hydrogen.	Space Tug
Solar Thermal				Hydrogen Storage, Hot Hydrogen.	Space Tug
Laser				Power Coupling, Ground Station Cost.	Rapid, Small Payload
Microwave				Power Coupling, Ground Station Cost.	Rapid, Small Payload
Gun Launch				High Gee Payloads, Aerothermal Loads.	Rapid, Robust Payload
Railgun				High Gee Payloads, Aerothermal Loads, Rails.	Robust, Small Payload
Space Platforms				Unfeasible.	---
Space Elevator				Materials, O, μmeteoroids, weather, vibrations...	Asteroid Mining
Breakthrough Physics				No known feasible concepts.	---

2.2 Microwave Augmentation of Solid Rocket Motors[16,17]

As identified above, one promising long-term technology for addressing the identified challenge for space access is beaming power (either lasers or microwaves) to a launch vehicle. A new operational concept was identified that represents the simplest possible implementation of beamed energy for launch: absorbing microwave energy that was beamed from the ground in the expanding section of the nozzle of a solid rocket motor as shown in Figure 3. This concept may allow BEP to be realized with only modest changes required to an existing vehicle. A small research effort was undertaken to determine the technological viability and potential long-term economic advantages of this new concept.

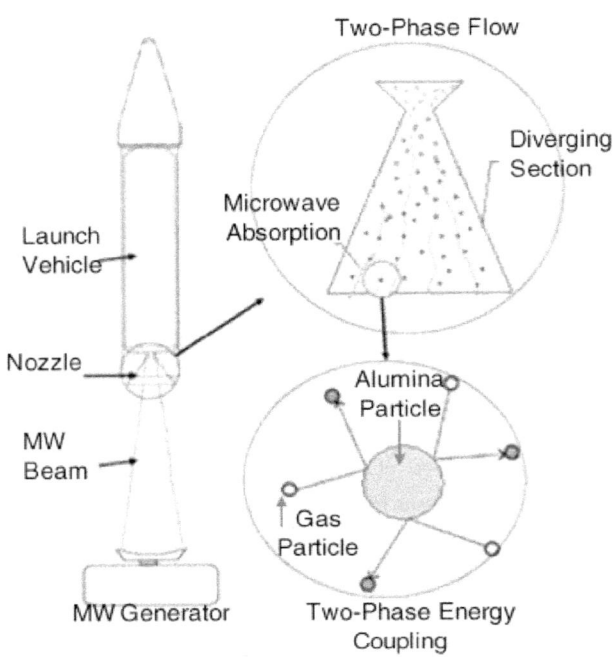

Figure 3. MW Thrust Augmentation Through Direct Coupling to Alumina Particles[16]

The first step in the research effort was to gain a general understanding of the technological viability of the concept. Specifically, the research effort aimed to determine: if the beamed energy could be effectively absorbed by the solid/molten particles in the exhaust, if the heated particles could then transfer their energy to the gaseous exhaust, and if the additional energy added to the exhaust was sufficient to measurably increase the performance of the stage. A relatively simple experiment was carried out to roughly determine the microwave absorption efficiency of alumina particles. A 2.45 GHz microwave source was used to illuminate a bed of powdered alumina (d = 10 μm) and the temperature of the particles was monitored using an infrared pyrometer while the total and reflected microwave power was measured. The particle bed was estimated to absorb 57% of the incoming microwave radiation. Two phase flow computational fluid dynamics CFD analysis was then used to determine the estimated effect on the exhaust flow of the absorbed microwave radiation in a representative motor (Castor 120). The simulations showed that a measureable increase in temperature and flow velocity could be achieved, but that most of the additional flow velocity was achieved in the first third of the expanding section of the nozzle.

A first cut design of both the ground station and the augmented vehicle was then assembled to gauge the potential performance increases and the projected cost and complexity of the complete system. The vehicle was chosen to be a Pegasus XL and it was assumed that the first two stages of the vehicle were augmented using microwave power. The power generation system (shown in Figure 4) consisted of an array of 665 total parabolic antennas, drawing 3.13 GW of input power in order to produce a nominal 1.0 GW at the launch vehicle at an altitude of 50 km. It was estimated that the total thrust produced by the motor could be increased by about 6% while the jet power would be increased by about 10% if a power density of 1 GW/m^2 could be delivered to the vehicle. It was concluded that this increase in performance probably did not justify the additional cost (several billion US dollars) of the ground station. Further advances in high power microwave sources and alternative applications of the ground station would be required to make this concept economically competitive.

Figure 4. Antenna Array and Thrust Augmentation[17]

3.0 ADVANCED CONCEPTS FOR NEAR-SPACE PROPULSION

Near-space is qualitatively defined as the range of Earth altitudes above which commercial aircraft can't produce sufficient lift for steady flight and below which the atmosphere is not rarefied enough for satellites to orbit with meaningful lifetimes. Near-space is often, but not always, quantitatively defined as the range of Earth altitudes from 20 km to the "edge of space", the Kárman line, at 100 km. There has been recent interest in long-duration near-space vehicles motivated by the possibility of operating with the relative benefits of both high-altitude aircraft and low-altitude satellites.[1] The common assumption is that operating in this range of altitudes will lead to certain advantages: persistent intelligence, surveillance, and reconnaissance (ISR), long endurance, beyond line of sight communication, and low cost access to space-like performance. There has been a wealth of recent reviews published on near-space systems and their potential performance levels.[2,3,4]

The problem is: it is very difficult to operate in the near-space atmosphere for long durations. The ambient gas pressure is too high for orbiting platforms, but too low for generating low-speed aerodynamic or buoyant lift (P_{amb} = 0.24 – 40 Torr). Some proposed aerodynamic or buoyant near-space platforms can operate at the very lowest end of the near-space altitude range, but are fundamentally incapable of going much higher (h < 40 km) while carrying meaningful payloads. Near-space is also cold (T_{amb} = 190 K to 270 K), contains significant atomic oxygen and ozone, experiences higher fluxes of UV than at Earth's surface, and can experience high wind speeds (often sustained >25 mph). In short, operating a long duration platform in near-space could lead to significant operational advantages, but the local environmental conditions make it extremely difficult. The Advanced Concepts Group identified one goal for near-space propulsion.

- *Demonstrate a new propulsion mechanism that allows sustained low-speed flight over the altitude range from 40 km to 100 km.*

3.1 Review of Advanced Concepts for Near-Space Propulsion[5]

A general near-space overview was conducted to: determine the relevant conditions in near-space, determine the required force levels for long duration platforms operating near-space, and evaluate potential thrust producing mechanisms for their force generation potential in the near-space environment. The attractiveness of operating in near-space lies primarily in the potential of achieving optimum performance per unit cost for ISR systems and beyond line-of-sight communication. Both atmosphere-based and space-based ISR systems encounter a fundamental trade-off between ground resolution and instantaneous area of access (total surface area available for a sensor) or footprint area (total surface area that a sensor observes at one time). As the altitude increases, both the instantaneous area of access and footprint area increase to include a larger fraction of the Earth's surface. Near-space systems also allow long term stationary observation (time over target is essentially unlimited). Near-space systems may also represent the optimal point in the trade-off between time on station and accessibility. Most of the proposed near-space vehicles can be landed, serviced, repaired and/or upgraded, and then relaunched. Near-space vehicles may, therefore, have the lifetime of a satellite (>10 years), but the accessibility of an aircraft.

What might be called "near-space craft," or "nearcraft," have actually existed since the 1930's. In 1934, a Soviet manned high-altitude balloon Osoaviakhim-1 flew (and crashed) after reaching an altitude of 22 km. However, the majority of the proposed near-term systems meant to endure in the near-space realm are designed to fly only in the lowest 10% of the near-space region: 20-25 km above the Earth's surface. Existing and near-term near-space systems are typically classified in three categories: free-floating balloons, dirigibles, and fixed-wing craft. All three have flown in near-space, but stratospheric dirigible flight has only been demonstrated for scaled platforms for very short time periods (hours) as of 2009.

An example of near-space balloons currently in use is Space Data Corporation's SkySite® Platform and its military equivalent, the StarFighter[TM].[6] The SkySite® Network is a constellation of balloons used to provide wireless data and communications coverage where there is none, or where existing coverage is cost prohibitive. Each balloon covers a range of more than 640 km, and carries a payload of less than 5.4 kg, at an altitude of 20-30.5 km, for 12-24 hours.

Fixed-wing craft offer many more options for long-term near-space travel, although they carry a considerably higher price tag, and require more resources and support. An example of a fixed wing vehicle that can fly in near-space is the Northrop Grumman RQ-4 Global Hawk, an unmanned aerial vehicle.[7] Its role is also surveillance, and it is capable of flying at an altitude of 20 km, with a speed of 650 km/h and an endurance of 36 hours. It is able to carry a payload of about 1,400 kg. Theoretically, the Global Hawk could provide persistent ISR with several aircraft in constant rotation, launched one and a half days apart.

As of early 2009, only two powered stratospheric airships have ever flown with a combined powered flight time of less than 4 hours. Aerostar International's HiSentinel airship represents the current state of the art of stratospheric airships that have actually achieved powered stratospheric flight.[8] In 2005, HiSentinel achieved 1.5 hours of powered flight at an altitude of 22.6 km with a payload of 27 kg.

All of the near-space vehicles described above are designed to fly in the lowest 10% of near-space. Operating near craft at altitudes above approximately 25-30 km in near-space requires either traditional platforms with extreme properties, or entirely new platforms with entirely new propulsion systems. Propellers for many proposed high-altitude airships already have a diameter of roughly 5m at an altitude of 20 km, and will scale larger with increased altitude.[9] Fixed-wing aircraft already have large wingspans at the low near-space limit; Helios had a wingspan of 75 m.[10] New technologies and materials will have to be developed to allow conventional vehicles to operate at these altitudes. Another potential solution, however, is to develop entirely new types of vehicles with entirely new types of propulsion systems.

The review concluded that forces on the order of 100N were required for expected near-space platforms. Different physical mechanisms were evaluated for their potential to achieve this level of force with dimensions, masses, and power consumption consistent with expected near-space platforms.

There is a wide range of potential air breathing electric propulsion mechanisms so a simple analysis was first used to define the desirable properties expected for application on near-space

11

platforms. The methodology was similar to the classical Isp optimization methods that minimize the total mass (including propellant and power supply) for in-space electric propulsion systems. Air breathing electric propulsion in near-space will not require stored propellant, but the thruster itself is likely to be large/heavy so a term representing it replaced the propellant term in the total mass expression. The total mass, m_t, is then given as a sum of the mass of the power supply, m_{ps}, and the mass of the propulsion system, m_{pr}, as given in Equation 1.

$$m_t = m_{ps} + m_{pr} = \alpha P_e + \beta A = \frac{\alpha g_o T I_{sp}}{2\eta} + \frac{\beta g_o^2 T}{\rho_{atm} I_{sp}^2} \tag{1}$$

Where α is the mass per unit power of the power supply, P_e is the power of the power supply, β is the mass per unit area of the air-breathing thruster, A is the open area of the thruster, g_o is the gravitational constant at the Earth's surface (g_o = 9.81 m/s^2), T is the thrust produced by the propulsion system, η is the thrust efficiency of the propulsion system, ρ_{atm} is the ambient air density, and Isp is the effective specific impulse for the propulsion system. Figure 5 shows typical total mass/thrust curves for different near-space altitudes for the conditions: α = 0.02 kg/W, β = 0.1 kg/m^2, η = 0.9.

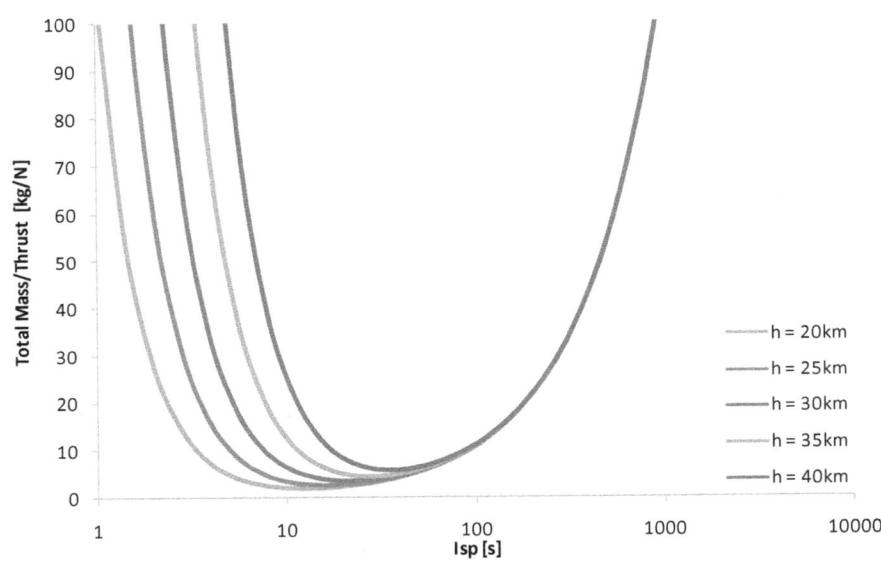

Figure 5. Total Mass/Thrust for a Range of Near-Space Altitudes

It is clear from the figure that the optimal condition (minimum mass) occurs at effective specific impulses that are much lower than for in-space propulsion systems. Many other parametric cases were studied and it was concluded that, over the entire range of expected parameters, the desired effective specific impulse was likely to be in the range of 10s of seconds versus 1000s of seconds for in-space propulsion. If an electric propulsion system were to be used to sustain long duration flight in near-space it would have to be large, lightweight, and operate at much lower specific impulse than traditional in-space versions. It must also operate with very low ionization fractions to achieve the required thrust efficiency at the desired low effective specific impulses.

One physical mechanism that seemed appealing under these considerations was electrohydrodynamics (EHD). In an EHD thruster a sharp object (such as a thin wire) is held at a high potential (10s of kV) relative to a ground plate. A corona discharge will occur near the wire to produce a stream of charged ions (both positive and negative can work). The ions are then accelerated in the electric field and slowly transfer energy to the surrounding neutral gas through collisions. The slow speed flow of the neutral gas is what produces the thrust. The review concluded that low levels of efficiency had been achieved until only recently, as of 2009, when demonstrated thrust efficiencies were approaching 7.5%.[11] The review also identified the relatively high ion mobility (2 x 10^{-4} m^2/Vs at sea level) to be a major limiting factor for achieving the required efficiency. It was determined that EHD propulsion system showed potential for long duration flight in near-space and a small level research effort was initiated to further study the concept (detailed later).

The mean free path of the ambient gas can range from 1 μm to 10 cm with altitudes ranging from 20 km to 100 km. This is approximately the scale of relevant features of a near-space platform indicating that rarefied gas effects should be considered for platforms operating at near-space altitude for long durations. The review concluded that the radiometric force, in particular, appeared to have the required properties for the propulsion system. The radiometric force occurs when a temperature difference is created between a hot and cold side of a plate or vane that is immersed in a rarefied gas. While the radiometric force was identified as early as the late 1700s it was still not understood to the level required to make general performance predictions. It was concluded that the radiometric force mechanism may have some potential for near-space platforms and a small research effort (detailed later) was initiated to: improve the physical understanding of the mechanism and make estimates of the possible performance of the mechanism.

The review concluded that operating a low-speed long-duration platform in near-space could yield revolutionary capabilities for both ISR and communication systems. New force production mechanisms were required to fly in most of the range of near-space altitudes and two were identified for further investigation: electrohydrodynamics and radiometric forces. Research efforts were initiated to investigate both mechanisms and they are summarized later.

3.2 Electrohydrodynamic (EHD) Propulsion in Near-Space[12]

A thorough review of potential mechanisms for producing thrust and/or lift forces at near-space altitudes concluded that the electrohydrodynamic mechanism showed some potential to meet the requirements defined for near-space electric propulsion systems: it uses low ionization degrees, yields effective specific impulses in the 10s of seconds, and can be made large and light-weight. In an EHD thruster a sharp object (such as a thin wire) is held at a high potential (10s of kV) relative to a ground plate. A corona discharge will occur near the wire to produce a stream of charged ions (either positive or negative will work). The ions are then accelerated in the electric field and slowly transfer energy to the surrounding neutral gas through repeated collisions. The slow speed flow of the neutral gas comprises most of the thrust in an EHD propulsion system.

There were two concerns identified for EHD propulsion systems with application to near-space platforms: their efficiency has historically been low (recent advancements had achieved

η_t = 7.5%)[11] and they typically produced low thrust levels. For this reason a small effort was conducted to model an ideal EHD propulsion system to determine its fundamental limits in performance (both thrust efficiency and thrust/area). A one-dimensional (1-D) model of an ideal EHD thruster was constructed and used to make estimates of the ultimate potential performance of the device. The schematic of the model geometry is shown in Figure 6 and the model is discussed in detail in the reference by Pekkar and Young.[12]

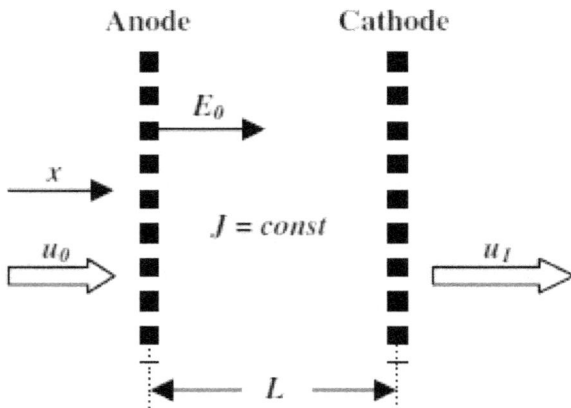

Figure 6. Simplified Geometry for 1-D EHD Analysis[12]

The maximum voltage applied across the gap between the electrodes is limited by the breakdown voltage of the working fluid (air) at the local conditions. The ion current in an ideal EHD thruster cannot exceed the space-charge-limited current; therefore, the thrust of such a thruster cannot exceed the thrust corresponding to this current. The model also concluded that the maximum thrust efficiency, η_t, of an EHD thruster can be effectively estimated using an analytical expression from Bonder and Bastien and given in Equation 2.[11]

$$\eta_t = \frac{1}{1 + \dfrac{u_D}{u_o}}$$

(2)

Where u_D is the ion drift velocity and u_o is the incoming neutral air velocity. The model showed that the thrust/area and thrust efficiency scaled in opposite directions for many parameters. For example, in the case of the applied voltage, it can be shown that increasing the applied voltage tends to increase the thrust/area, but it also tends to decrease the thrust efficiency as shown in Figure 7. The "nominal" lines are shown for typical parameters at sea level ($\mu_i = 2 \times 10^{-4}$ m²/Vs (ion mobility), $L = 0.1$ m (thruster length), and $u_o = 10$ m/s). It is shown that it may be possible to either achieve high thrust efficiency or relatively high thrust, but not both at the same operating condition. It must also be remembered that these results are achieved under optimistic assumptions and real performance levels are likely to be much lower.

The analysis identified the parameters involved in increasing the device performance: incoming flow velocity, decreasing the ion mobility, and decreasing the gap distance. Also plotted in Figure 7 are three additional sets of curves where each of these parameters has been adjusted sequentially. The next set of curves, labeled as "augmented" shows the increase in performance that would result from reducing the ion mobility by one order of magnitude ($\mu_i = 2\text{x}10^{-5}$ m^2/Vs). One potential method for achieving this decrease would be to apply an axial magnetic field to cause the ions to spiral; effectively increasing their collision cross-section for the direction of motion which results in a decrease in the ion mobility. As can be seen from the figure the thrust efficiency can be significantly improved while the thrust/area is only marginally improved by reducing the ion mobility by an order of magnitude. The third curve, labeled "increased flow" achieves similar additional improvements by increasing the incoming flow velocity from 10 m/s to 25 m/s. This may be achieved by using a lightweight collection scoop. The fourth set of curves is achieved by reducing the thickness from 0.1 m to 0.05 m. This has a profound effect on the thrust/area, but also causes a small reduction in the thrust efficiency. Optimization and additional improvements are required, but this simple analysis has already yielded potentially an order of magnitude increase in both thrust efficiency and thrust/area. Figure 8 shows similar curves for an EHD propulsion system at an altitude of 20 km. Similar trends are shown, but the performance levels have decreased due to the increase in ion mobility indicating that additional enhancements would be required for EHD propulsion systems to be useful in near-space.

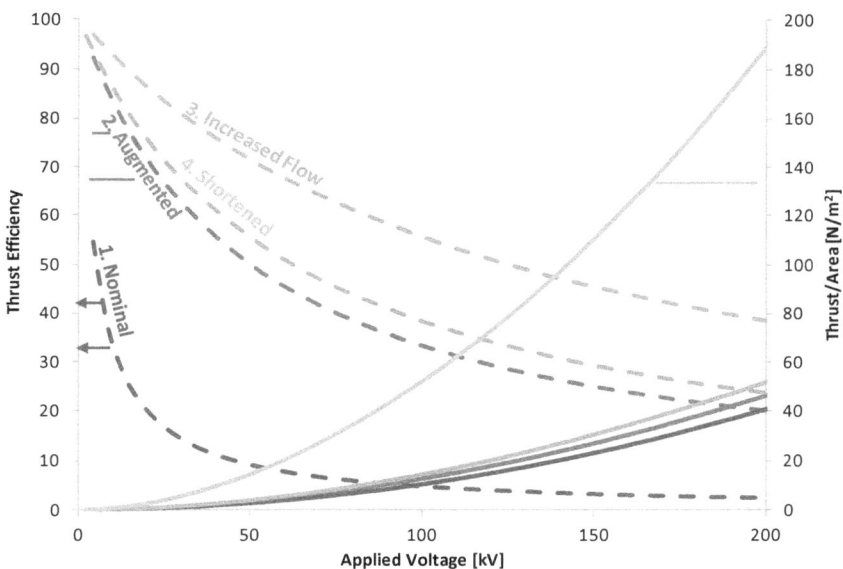

Figure 7. Optimum EHD Performance at Sea Level

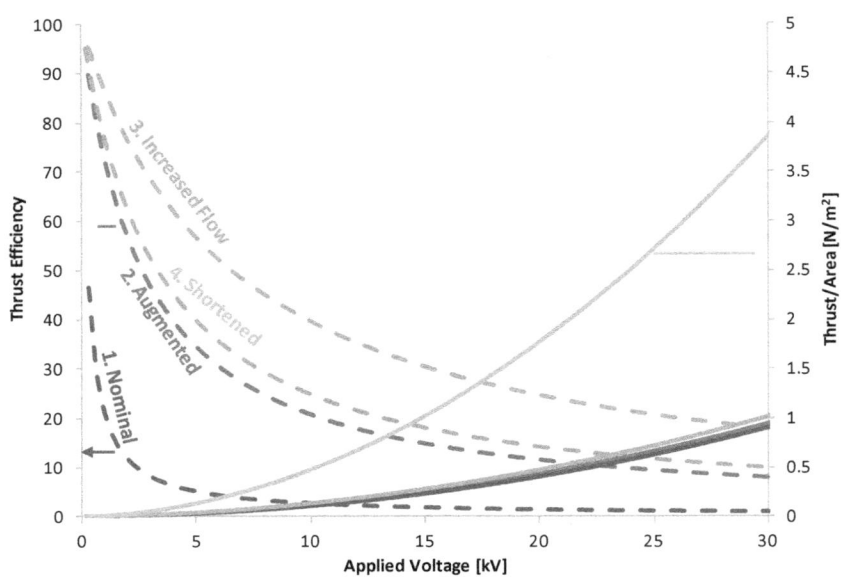

Figure 8. Optimum EHD Performance at h = 20 km

3.3 Radiometric Force Propulsion in Near-Space[13,14]

A thorough review of potential mechanisms for producing thrust and/or lift forces at near-space altitudes concluded that the radiometric force may be able to meet the requirements. The radiometric force is produced when a temperature difference is maintained across a vane or thin surface that is contained in a rarefied gas. The temperature difference is, often times, produced by illuminating one side of the vane. Accurate predictions of potential performance levels were impossible during the near-space review because a full understanding of the radiometric force mechanism did not exist. The radiometric force is actually composed of three component forces (area, edge, and shear) that act on the radiometer vane, as shown in Figure 9. The relative magnitude of each force was not known for arbitrary conditions and since they scale very differently there were large discrepancies in performance estimates based on which component force was assumed to be dominant. A **computational** and experimental effort conducted at University of Southern California (USC) and University of Colorado at Colorado Springs (UCCS) was used to: improve the basic understanding of the force, determine what factors had the largest effect on the force, and then use the improved understanding to design a potential near-space platform that could operate over a range of near-space altitudes.

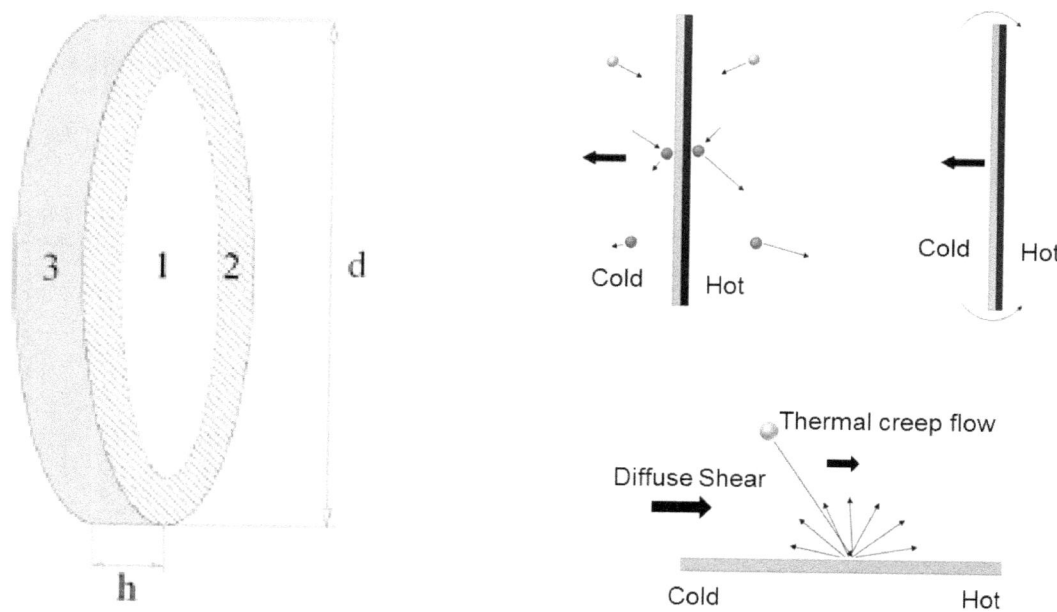

Figure 9. Area (1), Edge (2), and Shear (3) Forces on a Radiometer Vane.[13]

The radiometric force research effort attempted to gain a thorough understanding of the radiometric force by investigating each factor that affects the force individually. Only a brief summary of the results will be given in this report, but more detailed descriptions are given in the reference by Ketsdever.[13] The research effort first investigated the effect of the Knudsen number on the radiometric force. Figure 10 shows the measured force per temperature difference for a 5 cm by 1 cm vane for various gases and as a function of gas pressure. At very low gas pressures the force decreases due to the decrease in molecule-wall collisions and at very high pressures the force decreases as the mechanism disappears (there is no radiometric force in continuum flow). The radiometric force usually reaches a maximum at a Knudsen Number of approximately 0.1. There are two factors that cause differences in the radiometric force with gas species: differences in accommodation coefficient and difference in collision diameter. The gas molecules must accommodate with the surface for a vane to experience the radiometric force which is what causes helium, with noticeably lower accommodation than the other gases in the plot, to experience the lowest force at the free molecule (low pressure) limit. In the transitional regime helium has a much larger mean free path (due to its smaller molecular diameter) and experiences a higher force than the other gases.

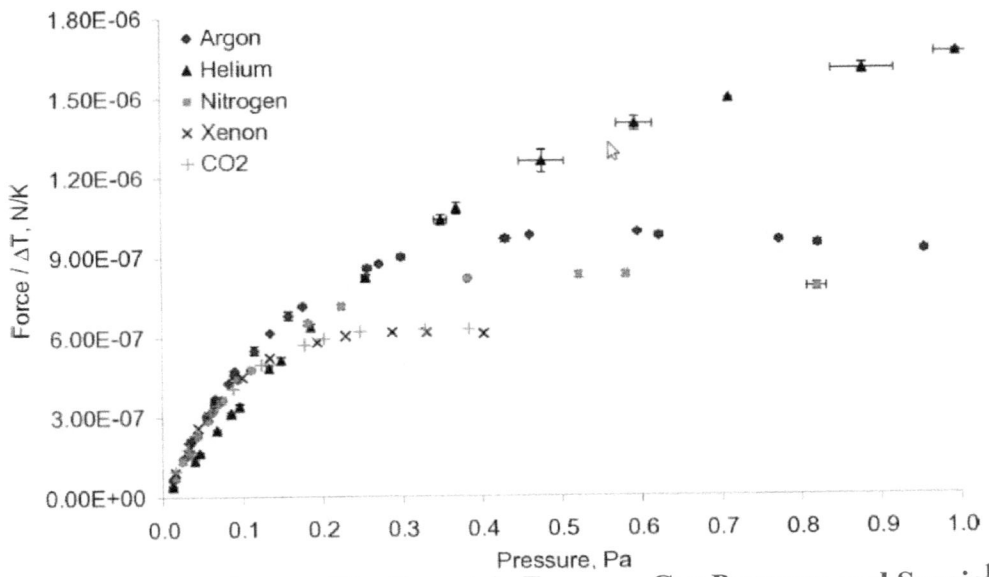

Figure 10. Dependence of Radiometric Force on Gas Pressure and Specie[13]

The radiometric force research then focused on determining the effect of the vane geometry on the radiometric force. Figure 11 shows the measured radiometric force for three different vane geometries: a small circle (d = 8.6 cm), a large circle (11.1 cm), and a large rectangle (7.62 cm x 12.7 cm). The large circle and rectangle had the same area, but different perimeters. The general conclusions from the shape investigations are that at high Knudsen numbers the radiometric force is predominantly an area force. As the Knudsen number decreases the edge force contributes more to the radiometric force. In general, small change to the shape of the vane (such as going from a circular to a square cross-section) will not significantly change the performance if the areas are the same.

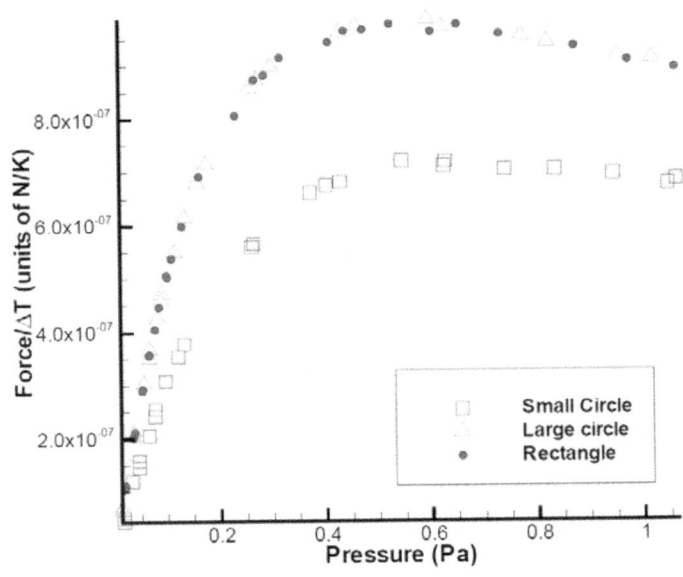

Figure 11. Dependence of Radiometric Force on Vane Geometry[13]

Vanes are not infinitely thin and the edges can experience a shear force which tends to decrease the force produced by the area and edge components. The research effort studied the effect of the vane thickness on the net radiometric force. Figure 12 shows the numerically predicted axial velocity and streamlines for radiometer vanes of two different thicknesses (conditions: 5.5 cm x 1.0 cm, T_h = 420 K, T_c = 395 K , P = 0.954 Pa). It is important to note that all of the numerical modeling for the radiometer effort was two-dimensional (2-D) due to computational limitations so comparisons with experiments are somewhat qualitative. The simulation showed that the vane thickness did have an effect on the flow produced by the radiometric effect.

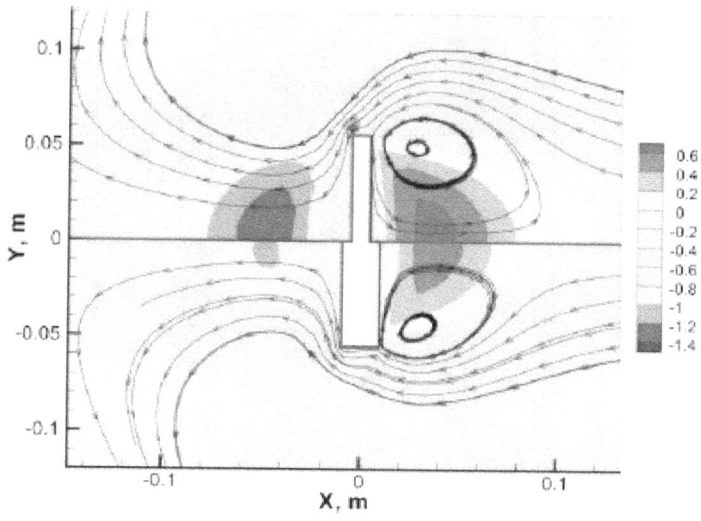

Figure 12. Axial Velocity and Streamlines for Different Vane Thicknesses[13]

A corresponding set of experiments were conducted to directly measure the effect of the vane thickness on the radiometric force. Figure 13 shows a picture of the "thick" vanes that were tested and a plot of the measured force as a function of thickness. It was shown that the thickness of the vane did not have a significant effect on the net radiometric force even when the thickness exceeded the characteristic length of the vane.

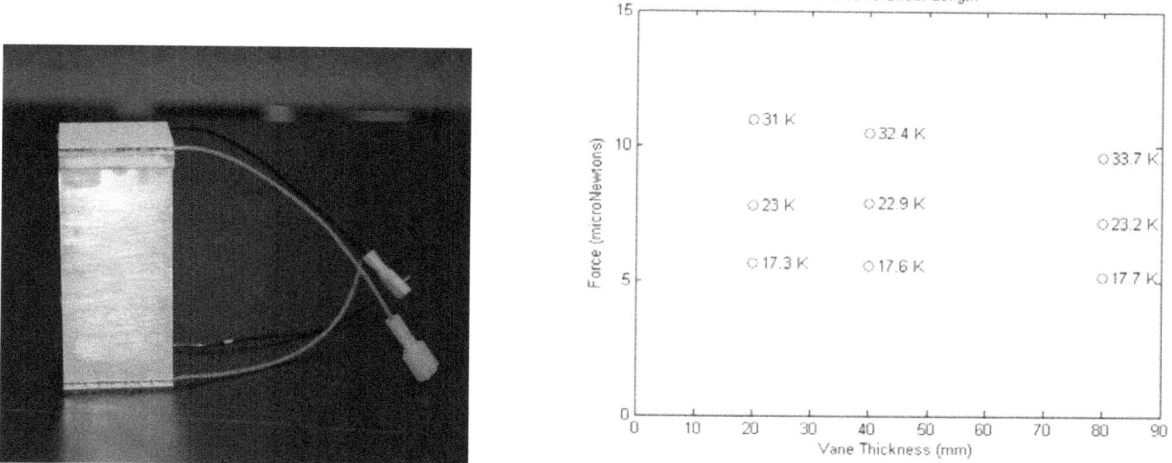

Figure 13. "Thick" Vanes and Their Effect on the Radiometric Force

A single radiometric vane typically produces only microNewtons of net force so a large number of vanes will be required to produce the force necessary for near-space platforms. The research effort also determined the effect of arranging vanes both beside each other and in front of each other. The flow field produced by an individual vane is very large compared to the vane dimension and so it was theorized that placing additional vanes nearby would significantly affect the flow. Figure 14 shows the numerically predicted streamlines and gas temperatures for a set of 4 vanes. The flow field varies significantly from the single vane flow field as shown in Figure 12.

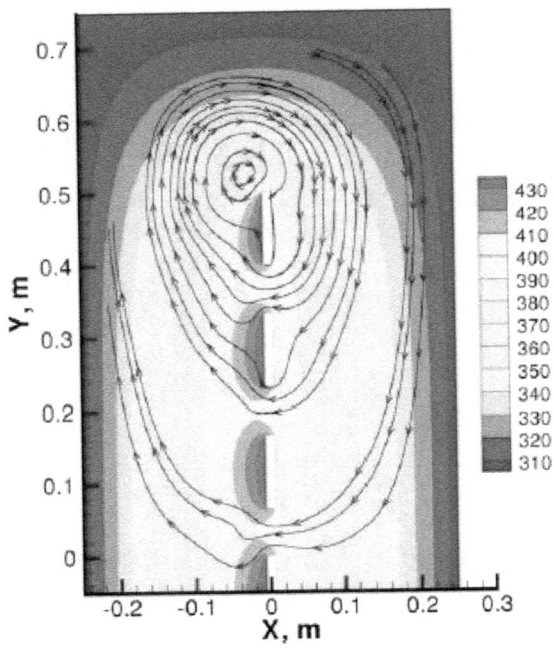

Figure 14. Gas Temperature and Streamlines for a Set of 4 Vanes[13]

A corresponding set of experiments were conducted to directly measure the effect of placing multiple radiometer vanes beside each other. In the experiments the gap, δ, was varied from a $\delta = 0$ (touching) to $\delta = 1$ (gap = vane length). Figure 15 shows the experimental configuration along with a plot of the normalized maximum force plotted as a function of gap separation. The force is normalized to the total of three individual vanes (effectively infinite δ). It was shown that the total radiometric force of three individual vanes can be reduced by approximately 20% if they are placed together. The force will then increase as the vanes are separated and by the time the separation reaches one vane length it will be roughly 95% of the total force produced by three individual free vanes.

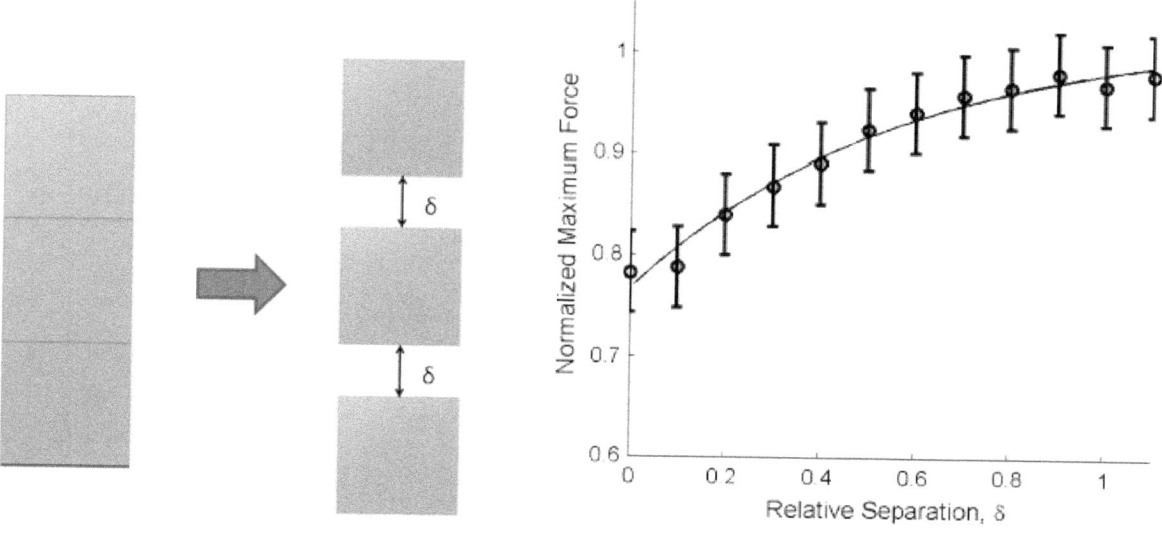

Figure 15. Effect of Placing Radiometric Vanes Beside Each Other[14]

A similar set of experiments were conducted to investigate the effect of placing radiometric vanes in front of each other as shown in Figure 16. Three individual vanes placed very far away would produce a total force of 24.5 μN for this configuration. The research showed that if the vanes were placed with a gap of more than roughly twice the characteristic vane length that the effect of the other vanes would be minimized.

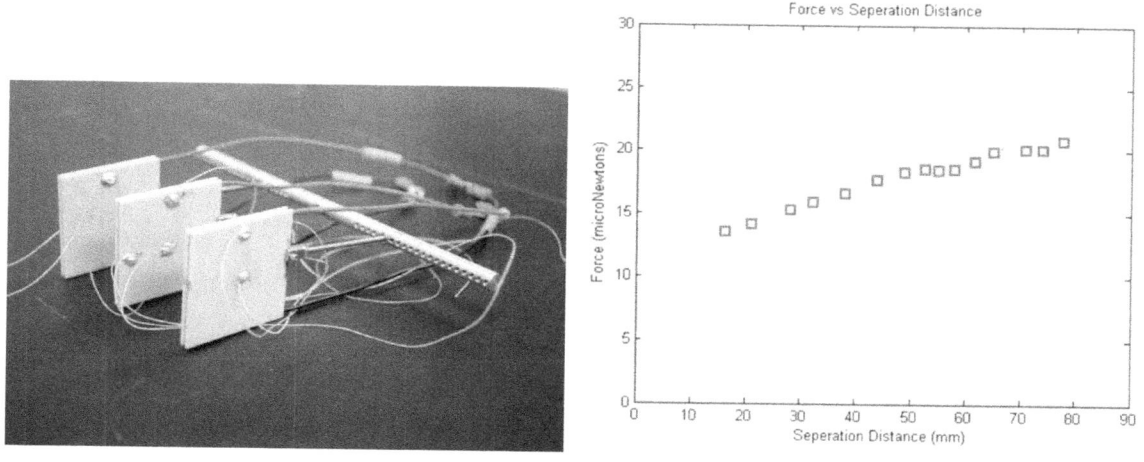

Figure 16. Effect of Placing Radiometric Vanes In Front of One Another

It was determined during the research that radiometric vanes produce flow fields that are significantly larger than the vane itself and that having boundaries near the vane altered the produced force. A research effort was conducted to determine the magnitude of this effect. Figure 17 shows the experimental arrangement used during the research and the measured force as a function of the proximity to the surface. It is shown that the force actually increases significantly (almost 40%) if the vane is placed right next to a surface. This effect will be very useful for the near-space applications discussed later.

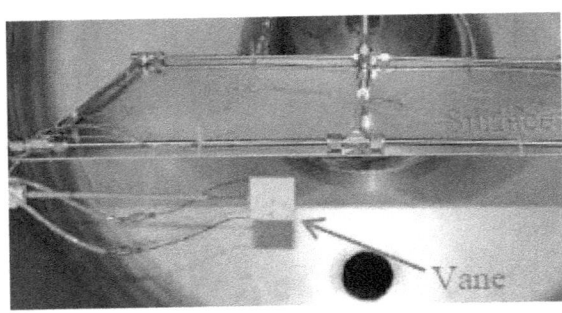
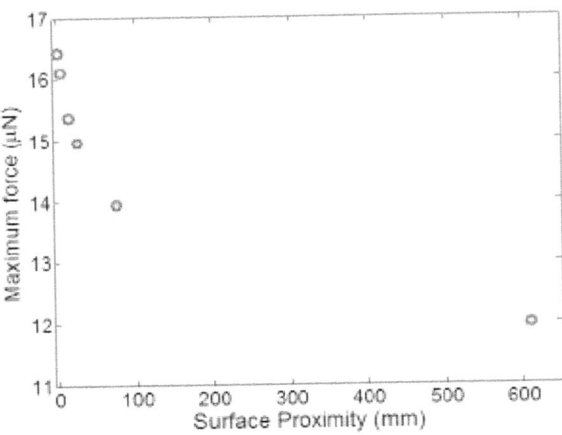

Figure 17. Effect of Placing Radiometric Vanes Near a Surface

Additional methods of potentially improving the performance of the radiometric vanes were also identified during the radiometric force research. One promising method, perforating the vanes with an array of small holes, was numerically predicted to yield improved force production. Future experimental efforts will explore the potential performance improvements possible for perforated radiometric membranes.

It is not yet possible to assemble all of the radiometric force data into a single expression useful to estimate the force for all practical conditions. An expression has been assembled, however, that does account for the variations due to the Knudsen Number, vane and gas temperatures, vane separation, and vane thicknesses. The radiometric force (F_{rad}) acting on a square vane with a side, a, and a thickness, w, that is part of a large radiometer array of identical vanes separated in the flow direction by a distance, s, and installed on a spacecraft surface that is in a low free stream velocity flow (v < 1m/s), may be evaluated as

$$F_{rad} = \frac{p}{2} A \cdot B \cdot C \cdot D \tag{3}$$

Here, p is the gas pressure. A is a vane shape dependent function given by

$$A = \begin{cases} a^2 & \lambda > \dfrac{a}{2} \\ 2\lambda(a - \lambda) & \dfrac{a}{10} \leq \lambda \leq \dfrac{a}{2} \\ 2\lambda(a - \lambda)\sqrt{10\lambda/a} & \lambda < \dfrac{a}{10} \end{cases} \tag{4}$$

22

Where λ is the air mean free path (70 nm at 1 atm, and roughly inversely proportional with pressure). B is a temperature gradient dependent factor given by

$$B = \sqrt{\frac{T_h + T_g}{T_g}} - \sqrt{\frac{T_c + T_g}{T_g}} \qquad (5)$$

Where T_h is the vane hot side temperature, T_c is the vane cold side temperature, and T_g is the free stream temperature. C is the vane separation dependent factor and is given by

$$C = \frac{2}{\pi} \arctan\left(\frac{2s}{a}\right) \qquad (6)$$

And D is the vane thickness dependent factor given by

$$D = \frac{1}{0.04\frac{w}{\lambda} + 1} \qquad (7)$$

The radiometer research effort improved the understanding of the radiometric force sufficiently that performance estimates could be made for near-space platforms employing the force. As a first step to designing a completely new platform based on radiometric forces a traditional dirigible shaped platform was fit with a radiometric thrust plate to determine its performance at a range of near-space altitudes as shown in Figure 18.

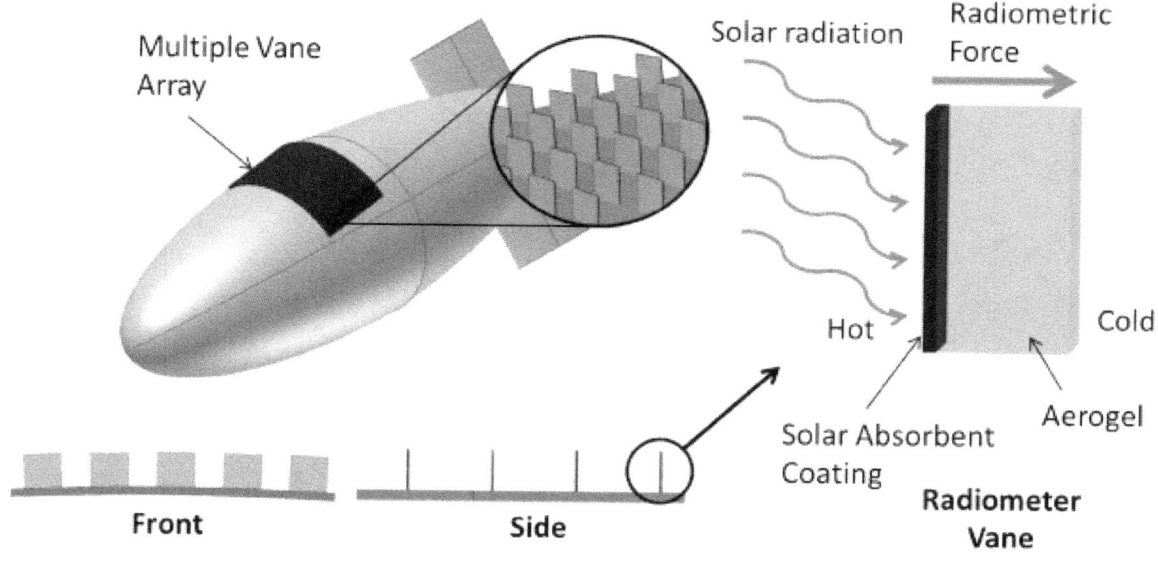

Figure 18. Radiometric Array Applied to Near-Space Dirigible Platforms[14]

The characteristics of the system designed for various altitudes are shown in the Table 3. Additional information about the system sizing can be found in the reference by Cornella.[14] It was shown that the chosen configuration could produce force levels consistent with expected requirements. The envisioned configurations are also consistent with the expected requirements for area and volume. The radiometric force plate looked most applicable at an altitude of 60 km which is of particular interest for the stated challenge. Future, much more thorough, designs including both a thrust *and* lift mechanism and a full, workable concept of operations is required to fully demonstrate the potential of radiometric forces in near-space, but the early indication is that they can provide useful amounts of thrust over the range of near-space altitudes of particular interest for the identified challenge.

Table 3. Propulsion-System Parameters for Various Operating Altitudes[14]

Parameters	Altitudes (km)				
	40	50	60	70	80
Pressure (Pa)	285.3	77.99	24.25	5.693	1.051
Vane Height (mm)	0.13	0.48	1.53	6.51	35.29
Vane Thickness (mm)	0.13	0.48	0.64	0.58	0.57
Vane Temperature Difference (K)	3.5	12.87	20	20	20
Number of Vanes/Area	9,491,928	727,920	88,290	5698	210
Thrust/Area (mN/m^2)	25.02	25.79	15.64	4.30	0.86
Areal Density (g/m^2)	5.62	8.63	12.50	13.94	14.90

4.0 ADVANCED CONCEPTS FOR IN-SPACE PROPULSION

The wide variety of satellite propulsion roles (orbital maneuvers, drag make-up, station keeping, attitude control) and the wide variety of potential space missions (even under the Advanced Concepts Group focus on Earth orbiting systems) lead to the definition of multiple potential in-space challenges. In general the ideal primary propulsion system for a typical satellite would operate very efficiently over the entire range of useful specific impulses (500s – 5000s), would be scalable to both very small and very large sizes, and would allow refueling (in particular with air) to allow a near infinite total mission ΔV. Assuming it was unlikely that all of the improvements required for the "ideal" propulsion system could be achieved in a single step from the current state-of-the-art, the strategy was chosen to address each aspect of the ideal propulsion system individually.

The most difficult specific impulse to achieve high thrust efficiencies (>50%) is around 1000s. There are flight qualified propulsion systems that can operate efficiently at specific impulses sufficiently below or above 1000s.[1] There are fundamental reasons it is difficult to design practical systems to operate at that condition. Electrothermal thrusters such as resistojets and arcjets can achieve high efficiencies at low specific impulses, but are fundamentally limited due to frozen flow losses and boundary layer effects as the specific impulse approaches 1000s. Hydrogen arcjets could potential achieve the required performance, but lack a compact hydrogen storage solution and are not space qualified. Both traditional electrostatic and electromagnetic propulsion systems can achieve high efficiencies at specific impulses significantly higher than 1000s, but cannot efficiently operate at 1000s due primarily to the ionization costs. Colloid thrusters can operate efficiently at 1000s, but there has been difficulty in practically scaling them up to useful sizes. One challenge identified for in-space primary propulsion systems is to "demonstrate a practical propulsion system that can operate efficiently (η>50%) at a specific impulse of 1000s." Several research projects were aimed at addressing this challenge: mini helicon experiment, capillary discharge experiment, and the macron experiment. Each research experiment is discussed in their individual sections below.

The information gained researching efficient thruster operation at a specific impulse of 1000s will be used to address the next challenge in developing the "ultimate" primary satellite propulsion system: "demonstrate a satellite propulsion system that achieves high thrust efficiency over the entire range of specific impulses from 500s to 5000s." This challenge was not addressed during the Advanced Concepts research and is left for future research.

Any satellite propulsion system must make a trade-off between being responsive or achieving a high total mission ΔV. Chemical propulsion systems are the most responsive and are capable of delivering their entire mission ΔV in hours. Electric propulsion systems are power limited and must decide between high thrust or high specific impulse. Even proposed propellantless systems such as solar sails and electrodynamic tethers have very efficient "effective propellant utilization," but are very slow to respond. Responsive spacecraft that can be refueled could significantly increase their total mission ΔV, while still maintaining their responsiveness. Spacecraft that can refuel by efficiently collecting ambient atmospheric gas would not require additional launches and represent perhaps the holy grail of achieving high thrust and high total mission ΔV. Air collecting/breathing spacecraft propulsion systems have long been identified as a truly revolutionary technology if they could be realized.[2-4] There are significant challenges with all aspects of such a system (propellant collection, compression and storage, and propulsion). Another identified in-space problem is then to "individually demonstrate the components that could enable ambient gas to be utilized as propellant for low altitude orbiting systems." No promising concepts were identified during the Advanced Concepts research effort and so there were no significant efforts towards meeting this challenge.

One apparent satellite trend is towards packing useful capabilities into ever smaller satellites. Satellites as small as 1 kg (cubesats) have been flown, although with very limited capabilities. Microsatellites, with masses up to 100 kg, that have some capabilities have also flown, but they have had limited onboard propulsion. A review was conducted to determine the existing state of the art for providing useful propulsion capabilities (both a high total mission ΔV and a high thrust) for a microsatellite.[5] It was concluded that "a satellite propulsion system that could provide a total mission ΔV of 1.0 km/s with a response time of days" would represent revolutionary improvement over the existing state of the art. Such a system would allow microsatellites to be delivered as secondary payloads, but still perform useful missions. This was defined as one of the in-space challenges for the Advanced Concepts research effort. It was also shown that bi-modal solar thermal propulsion systems (that included thermal energy storage) were a promising candidate solution. The High Energy Advanced Thermal Storage (HEATS) project was initiated to investigate the potential of advanced thermal storage solutions for such a bimodal solar thermal system and will be summarized below.

A completely contradictory trend also exists, however, where ever larger and more powerful satellites are continuously being launched, particularly for communications. It is not unreasonable to discuss spacecraft with power levels of 100 kW to 1 MW within the Advanced Concepts time-frame (15-50 years). Recent reviews of potential future high power electric propulsion systems have demonstrated that there are multiple potential options for this power level.[1] No Advanced Concepts challenges were defined for this area, but one concepts was researched to determine its potential performance in this area and scaled to other areas, field reverse configuration (FRC) propulsion.

Nontraditional satellite operations may also be envisioned in the Advanced Concepts timeframe. One space operations concept currently being explored is the fractionated satellite. In the concept, a large cluster of small satellites are flown in a platoon rather than using a single larger satellite. Such concepts may allow larger effective apertures for viewing along with more robustness against failure. One difficulty with such systems is maintaining the required separation without consuming prohibitive amounts of propellant. Without a constant force acting, all formation flying satellites are in orbits that contain the center of the Earth. Their orbits cross twice every period, and the satellites tend to converge on each other unless a continuous separation force counteracts this convergence. The magnitude of the required separation force depends on the total mass of the two spacecraft and their separation distance. In low Earth orbit, thrust forces between 100 and 1000 mN are required for satellites pairs with masses between 100 and 1000 kg separated by 1 km. Another difficulty for such systems is avoiding thruster plume contamination of the satellites. In general, this operational concept requires a propulsion system that achieves a very high effective specific impulse and must have a well behaved plume. Another problem for in-space propulsion is to "demonstrate a propulsion system that can allow satellite formation flying at significantly higher effective specific impulse ($Isp_{eff} > 100,000s$) and avoid contamination of other satellites in the constellation." The liquid droplet thruster concept was identified as a potential solution to this challenge and the corresponding research effort will be summarized below.

A Near Earth Object (NEO) Threat Mitigation Study was also conducted in an effort to define the needs for advanced propulsion. The results of this study will also be detailed below.

The list of key challenges identified for in-space systems are summarized as:

- *Demonstrate a practical propulsion system that can operate efficiently ($\eta > 50\%$) at a specific impulse of 1000s.*
- *Demonstrate a satellite propulsion system that achieves high thrust efficiency over the entire range of specific impulses from 500s to 5000s.*
- *Individually demonstrate the components that could enable ambient gas to be utilized as propellant for low altitude orbiting systems.*
- *Demonstrate a microsatellite satellite propulsion system that could provide a total mission ΔV of 1.0 km/s with a response time of days.*
- *Demonstrate a propulsion system that can allow satellite formation flying at significantly higher effective specific impulse ($Isp_{eff} > 100,000s$) and avoid contamination of other satellites in the constellation.*

4.1 Overview of High Thrust Microsatellite Propulsion Systems[5]

Microsatellites have been suggested as a means of enhancing a variety of proposed space missions, ranging from low Earth orbit to Solar System exploration. Light-weight (100 kg class and smaller) microsatellites, combined with miniaturized spacecraft components, are a well-established technology proven to reduce the costs and enhance the capabilities of certain space missions. The relatively small mass of microsatellites could allow for drastically reduced launch costs; reduced development times for microsatellites may also result in the use of more modern technology, which can enhance capabilities and mitigate some of the compromises made to reduce system mass. Additionally, the concept of producing several similar or identical microsatellites provides another avenue for targeting the needs of specific missions. A microsatellite can also be constructed and launched (i.e., to meet a specific need or to replace a failing satellite) on shorter notice, and likely with decreased expense, when compared to the process required for a full-sized satellite.

The above-listed advantages of microsatellite missions are well known, but as yet, microsatellite capabilities in terms of the ultimate velocity increment (ΔV) available for station keeping, orbit transfers, and other maneuvers have been viewed as somewhat limited. The goal of this project was to survey the propulsion technologies available for microsatellites and to evaluate each technology for potential use in a demanding mission requiring high thrust and high ΔV (i.e., greater than 1 km/s). Miniaturized propulsion systems that were analyzed based on published data included bipropellant microrockets, monopropellant microrockets, cold gas thrusters, solar thermal thrusters, electrothermal microthrusters, electrostatic thrusters, and others as shown in Table 4. The references for each individual technology are given in Scharfe and Ketsdever.[5]

Table 4. Summary of Propulsion Technologies Available for Microsatellites[5]

Thruster Type	References	Thrust	I_{sp} [s]	Power [W]	Thruster Mass
Hall/Ion	11,22-24,26-34	0.4-20 mN	300-3700	14-300	$\leq$ 1 kg
FEEP/colloid	11,25,26	0.1 μN-1.5 mN	450-9000	1-100	0.1-1 kg
Electromagnetic	11,22,26,35	0.03-2 mN	200-4000	$\leq$ 10	0.06-0.5 kg
Electrothermal	6,11,26,36-39	$\leq$ 220 mN	50-250	3-300	0.1-1 kg
Cold Gas	16,26,40	0.5 mN-3 N	40-80	-	0.01-1 kg
Monopropellant	4,11,25,26,41-43	1 μN-1.5 N	100-230	$\leq$ 6	0.01-0.5 kg
Bipropellant	4,6,11,16,19,25,26,44	1 μN-45 N	100-320	$\leq$ 6	0.01-0.5 kg
Decomposing Solid	45		230		
Laser Micro. (ablation)	18	1 μN	100-300	2	
Laser Micro. (ignition)	18	1-10 mN	37-100	-	
Laser Plasma	3,46	0.1-1 mN	500-1000	2	$\leq$ 1 kg
Hollow Cathode	47	1 μN-10 mN	50-1200	5-1000	
Solar Thermal	4,8,9,20,48-50	56 mN - 1 N	200-1100	-	$\leq$ 10 kg

It was found that existing bipropellant microrocket designs provide a high thrust value, combined with a 300 second specific impulse, allowing for response times of only a few hours for such a mission with ΔV requirements over 1 km/s. Miniaturized electrostatic thrusters provide the largest ultimate ΔV capability, approaching 10 km/s, but with a very low thrust level and therefore a response time capability of several months or years to achieve 1 km/s velocity change. Newly developed micro-solar thermal systems fill in the middle ground of these two options, providing the moderate thrust levels and specific impulse values necessary for a response time on the order of one day and a ΔV of several km/s. In general satellite usage, solar thermal propulsion has several drawbacks, including the mass required for solar light collection and concentration, and the requirement for solar illumination to provide thrust.

Ultimately, for the Department of Defense and other parties that may be interested in the type of aggressive microsatellite mission hypothesized in this work, a relatively fast response time will likely be a strong deciding factor in selecting the propulsion technology. It was therefore concluded that the miniaturized chemical rockets were a strong candidate technology, but were yet to be exhaustively tested to precisely measure critical characteristics such as thrust, total propellant throughput, and thruster lifetime. Proposed solar thermal propulsion systems can exceed the *Isp* of a chemical system, with a relatively small sacrifice in system thrust and response time. While the robustness and reliability of such a simple system is likely high, the development of these thrusters specifically for microsatellites, including options such as fiber-optic coupling and thermal storage, appears to be at a relatively early stage of development. Further research and development of this technology is recommended. The review is discussed in more detail in Scharfe and Ketsdever.[5]

4.2 NEO Threat Mitigation Studies[6]

It is well known that NEO and other celestial bodies can be a threat to human existence and civilization. While impacts with large objects occur with very low probability, their consequences can be so catastrophic and irremediable that a program to alleviate this type of threat would seem a very prudent decision. Recently (as of 2006), NASA has been tasked with detecting and characterizing NEOs. However, the role of mitigating these threats is yet to be defined, and may be suitable for USAF responsibility. Mitigation approaches are varied and require further study, but of particular concern are the most difficult scenarios of interception, involving objects with large mass and little advance warning. Although threat mitigation will require important decisions, authorizations, multi-agency coordination, and likely international collaboration, some essential long-term planning steps are required to develop and mature key technologies in order to defeat these threats. These steps can be part of an overall long-term strategy for space exploration and utilization that can be part of a global peace-time Department of Defense (DOD) activity, and that can also greatly increase the welfare of mankind.

The issue of threats of impacts from Near Earth Objects and other celestial bodies is a topic which has been addressed many times before and will very likely continue to be so. The NEO Threat Mitigation report was motivated by the release of the draft Final Report on Near Earth Object (NEO) study[7], prepared by NASA for the US Congress in December 2006; this document has been labeled by NASA as "pre-decisional" and only a draft – a status that has not yet changed by the date of the NEO Threat Mitigation Report (October 2007). In 2005, the US Congress directed NASA to perform an analysis of alternatives to "detect, [...] characterize potentially hazardous objects (PHO), and submit an analysis of alternatives for threat mitigation". That 275-page document contains the summary of several years of studies performed by multiple investigators and published or presented elsewhere.[7] As such, the NASA report is an excellent summary introduction to the overall challenge presented by NEO threats, feasible technical approaches, and logistical and budgetary consequences. The NEO Threat Mitigation Report focused on the potential role of USAF and the relationship with advanced technology development programs that are sponsored by the USAF or are candidate for such sponsorship. While there can be many geo-political implications of the threat itself and approaches to its remediation, the report did not attempt to discuss these in great detail; they were mentioned only as contextual elements of the technical discussion. It is also important to emphasize that the report does not represent a complete analysis of the technical requirements, nor does it describe an official position by the USAF; there are currently too many unknowns preventing such decisive statements and important decisions involving Planetary Defense can only be made at a high level of authority.

The report first defined the threat posed by the impact of asteroids and comets with Earth. It is noted that the threat of collision decreases with increasing object size, but the consequence of collision greatly increases. The threats from NEOs were divided into three classes as shown in Table 5. The primary concern is the threats due to very large NEOs that have the potential to end civilizations or cause species to go extinct.

Table 5. NEO Threat Classes[6]

Threat Class	Diameter (m)	Threat
Small	d < 140	Mostly Inconsequential, Not Worth Addressing
Medium	140 < d < 1000	Above Threshold of Concern, Regional Event
Large	d > 1000	Major Threat to Survival of Civilization or Species

The entire mitigation process was evaluated including: detection and tracking, object characterization, and mitigation. In general, the limited options that were identified were insufficient to address the NEO threat and a significant developmental effort would be required to enable any mitigation solution. It is sobering to realize that we have no capability at the present time to deal with the threat of a comet-like impact; should there be such an object discovered nothing could be done to avert the resulting extinction-class event – despite misconceptions popularized by the entertainment industry. For this class of impulse requirements, only buried nuclear explosives could be effective, but we currently do not have suitable launcher and space vehicle systems capable of intercepting the object. If the threat is a solid-core NEO, we do not have the capability to drill into the object. If we attempt a nuclear stand-off deflection, we do not have nuclear devices with sufficient yield…It would therefore seem very prudent to develop the required capabilities as soon as possible. The threat from smaller objects can then be handled by the same capability, while more cost-effective approaches can be designed and demonstrated throughout a longer-term campaign. Fortunately, a significant amount of technology being developed principally for the DOD and NASA can be leveraged as a starting point in such a campaign. If humanity or national survival are not a sufficient rationale, the concern over "wasting" resources for a Planetary Defense program can be greatly reduced when considering the impact of the needed concepts and technologies required for advanced DOD missions and in the long-term, for space colonization and exploitation. A coherent, long-term strategic plan must be decided with utmost urgency, for the completion of various critical phases of a Planetary Defense program. Some key first steps include the following:

- Need to (at least) stockpile current multi-megaton nuclear weapons, and preferably design and develop much higher yield devices, such as 3-stage weapons (100 Mt and above). These weapons are obviously not in stock and the designs should be optimized for radiative energy transfer (standoff deflection), until the technology to drill into various asteroid and comet cores is developed, in which case the 10 Mt-class may be sufficient. This would of course be highly controversial and raises the issue of testing. Given that the desired yield for stand-off deflection for comet-like impacts is much larger than the largest weapon tested by the US, it seems prudent not to rely exclusively on numerical predictions; however, such weapons could not be tested on Earth, even under-ground, and in-space testing would require a revision of international treaties.
- Need for very reliable heavy-lift launch capability, beyond heavy Delta IV capability. The ARES-V launcher being developed by NASA may provide just enough payload capability to launch a nuclear weapon of about 200 Mt (mass of 80 tons). To obtain much higher yields for a stand-off deflection of a comet-like impact (extinction-class, little warning), multiple launches would be required. The risk of failure would increase correspondingly, and the launchers may not be available in time. Thus, there is also a critical need for rapid manufacturing, assembly and stockpiling of launcher components; this requirement matches a large-scale and long-term strategy for future DOD missions and space utilization.
- Need to develop multi-MW nuclear power, with multi-year (>10) lifetime; this is critical to multiple applications, from Planetary Defense, DOD missions and space colonization. Such technology is significantly more advanced than prior efforts at deploying space nuclear power, notably for efficiency (target: >50%) and mass efficiency (α <5 kg/kWe). Yet a consistent, long-term research and development (R&D) effort can bring the technology

towards these challenging figures of merits, and make nuclear space power a formidably attractive technology option.

- Need to develop MWe-class high Isp, high-thrust propulsion systems. This type of R&D is currently under-way and needs to be accelerated; however, its true potential will not be realized until the power source (item above) can be developed as well.
- Need to develop capability to deploy very large (>1 km) structures: radar antennas, "nets", solar collectors, and beam power antennas. The latter are especially useful as a component of a space infrastructure, allowing innovative concepts for challenging DOD missions, reducing launch costs and allowing the expansion of human presence in space. These large-scale structures will need to be assembled by advanced robotic operations and artificial intelligence. In the long-term such complex, large-scale structures can be used to exploit the natural resources within the solar system.
- Need to improve the longevity of spacecraft. This is required because of the long-duration missions for observation, tracking and deflection of the NEOs. Advanced materials, shielding concepts and embedded multi-functional structures are required, but this capability can be also augmented by redundancy and self-repairing, or by the deployment of repair/refurbishing stations.

Thus, Planetary Defense presents some unique operational and technical challenges. Yet when considering the context of a long-term space infrastructure development, the requirements for a successful Planetary Defense campaign provide multiple opportunities to bring forward the concepts and technologies that will immediately impact DOD missions across the national security spectrum, and allow future long-term commercial and societal benefits. The DOD can play an important leading role in the initial design and implementation of this long-term strategy, leveraging other key agencies such as Department of Energy (DOE) and NASA. Other leveraging opportunities are likely to be found as the strategy matures. It is clear that other nations have ambitious long-term goals to implement such strategies. While this may be considered as an opportunity for international collaboration, it also implies that the US must take important steps forward as soon as possible if it wants to remain competitive, or even be allowed to play a major role in such international endeavors.

4.3 Mini Helicon Thruster (mHTX)[8]

As discussed earlier, the Advanced Concepts Group determined that a propulsion system that could operate efficiently ($\eta_t > 50\%$) at intermediate specific impulses (Isp ~ 1000s) would yield a new capability for highly responsive satellites and would represent a practical first step in the development of an "ideal" thruster that could operate efficiently over a wide range of specific impulses (500s - 5,000s). More specifically: *there is a need for efficient thrusters operating in the ~10-1000W power range producing 0.1-100mN, and not relying on the expensive and increasingly rare xenon gas as a propellant source.* Early in the Advanced Concepts project, Dr. Cambier identified the helicon discharge as an interesting pulsed high-density plasma source and one concept that could potentially address this issue.

An experimental study of a mHTX for space propulsion was performed during the period 01/2005-02/2008 at the Massachusetts Institute of Technology (MIT) Space Propulsion Laboratory (SPL), under a contract from Air Force Research Laboratory (AFRL) at Edwards Air

Force Base (Program Manager: Dr. J.-L. Cambier). With Dr. O. Batishchev as a Principal Investigator, twelve MIT students from three MIT departments were involved into this project, resulting in one Ph.D. and five M.S. theses.

The initial step of the experimental program consisted of developing a working device capable of operating in a small vacuum facility with a limited pumping capacity. Several configurations of the radio-frequency (RF)-delivery sub-system were tried prior to achieving the first inductively coupled discharge. Low-current electromagnets were built to excite the helicon mode in Argon. The efforts were made to gain the required controllability of the discharge parameters. Various RF antenna designs, materials and connections were tested. Preliminary gas and plasma characterization was performed using borrowed instruments from AFRL/Hanscom.

The second step included detailed characterization of the mini-helicon operation. A high resolution UV-VIS system for spectral analysis was designed and built using Air Force Office of Scientific Research (AFOSR) support (Program Manager: Dr. M. Birkan) via Defense University Research Instrumentation Program (DURIP) grants. A new, fully computerized system to control RF-power, magnetic field strength and gas flow was implemented. The spectral sub-system was separated from the rest to improve stability. Stable operation with argon (Ar) and nitrogen (N_2) gases was achieved in double- and single-magnet configurations. Invasive Mach probes were machined and used to measure the plasma exhaust.

The third step included system optimization and development of the refined diagnostics to quantify energy balance, heat fluxes, plume detachment and divergence, thrust force, etc. A half-year delay due to serial failure of the SPL's vacuum system components forced a correction to the schedule. The RF-delivery and matching networks were re-designed to minimize losses and gain control. Compact high-current electromagnets were built. Seeding light impurities was attempted to facilitate diagnostics. In-situ retarding potential analyzer (RPA) probing and noninvasive Doppler shift measurements has indicated high plasma acceleration in the mHTX with a collimated plume formation. Variability of the specific impulse by controlling power density was demonstrated.

The helicon discharge was designed and built in several different configurations using low- and high-current electromagnets. The program was designed to address several objectives: (i) to achieve a high coupling of RF-power to plasma and efficient acceleration to >20 km/s; (ii) to evaluate the scaling relations between power density, flow rates and B-field strength; (iii) to demonstrate operation with common molecular and atomic gases, and mixtures; (iv) to provide detailed characterization of the performance using spectroscopic and invasive techniques; (v) to experimentally demonstrate variable specific impulse capability. These objectives were successfully achieved. Stable discharges with well-collimated plume exhaust were achieved for mixtures of Ar and N_2, and air. The parameters of the plasmas were measured using high-resolution spectroscopic and in-situ diagnostics, including super-fine emission line structure and isotopic shift of a seeded impurity. Plasma plume flow velocities in the 10-40 km/s interval were measured, and plume acceleration could be easily varied by altering applied RF-power or propellant flow rate.

Very high ~4 MW/m^2 power density throughput was achieved in a steady-state regime. Operation with permanent magnets was attempted as well, including a magnetic nozzle and the possibility of a double layer formation. The experimental program thus demonstrated the potential of the mini-helicon discharge as an excellent space thruster design, given its high throughput, absence of internal electrodes and cathodes, potentially long life time, very low divergence of the plasma plume, flexible choice of propellants, and variable specific impulse. Basically, this project delivered a brand new electro-thermal propulsion concept with compact design and controllable specific impulse, capable of operating with diatomic and noble gases. The concept has favorable scalability to both lower- and high-powers compared to known EP devices.

Additional details on this project are available in the technical report: AFRL-RZ-ED-TR-2009-0020.[8]

4.4 Capillary Discharge[9,10,11,12]

As discussed earlier, the Advanced Concepts Group determined that a propulsion system that could operate efficiently (η_t > 50%) at intermediate specific impulses (Isp ~ 1000s) would yield a new capability for highly responsive satellites and would represent a practical first step in the development of an "ideal" thruster that could operate efficiently over a wide range of specific impulses (500s - 5,000s). Early in the Advanced Concepts project, Dr. Cambier identified the capillary discharge as an interesting pulsed high-density plasma source and one concept that could potentially address this issue. In the 1980s Burton explored a similar concept, a pulsed electrothermal thruster, which had promising levels of performance, but which still required additional research to show proof-of-concept as a system.[13]

A capillary discharge is a pulsed electrothermal discharge that takes place in a long narrow tube made from an ablating material (traditionally polyethylene) as shown in Figure 19. A pulse forming network or similar pulsed circuit is used to apply a voltage waveform to the discharge. Once ignited, the potential difference (due to the pulsed circuit) across the electrodes at the ends of the tube maintains a current through the tube. The plasma in the tube radiates which is efficiently absorbed in the tube wall due to the high solid angle filled by the tube for most locations. The absorbed radiation leads to ablation of the tube material from both direct photoablation and thermal pyrolysis. The ablated gas is then both dissociated and partially ionized before entering the bulk plasma core. The hot gas/plasma is also continuously being expelled out of an expanding nozzle at the end of the capillary to produce thrust.

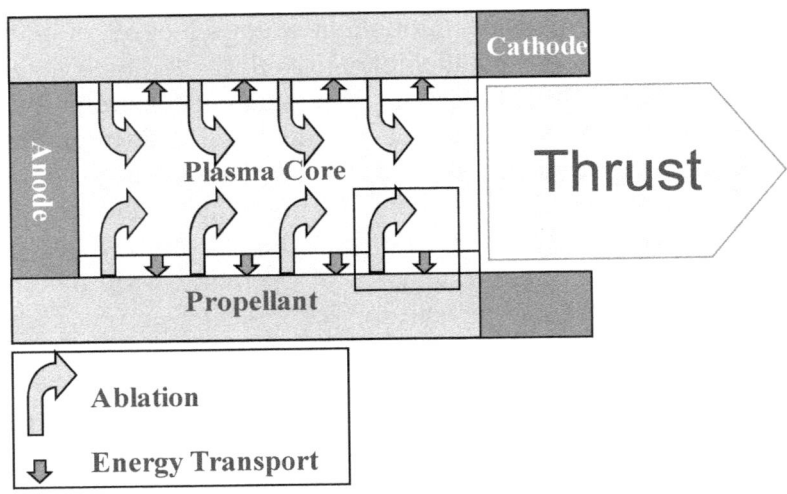

Figure 19. Physical Operation of a Capillary Discharge

There are several basic factors that lead to the expectation of efficient plasma generation and thrust production in capillary discharges. One factor is the efficient radiation absorption and energy use (ablation, dissociation, ionization) at the capillary walls. The majority of the energy loss from the plasma core of capillary discharge plasma is in the form of radiation. The capillaries are typically long, narrow cylinders so photons radiated in nearly all directions from most internal points will strike the capillary wall and not be lost to the surroundings. During the Advanced Concepts research it was also shown that it is important that the material is highly absorbing (high extinction coefficient) to ensure that the radiation is absorbed very near the capillary wall surface for efficient ablation.[11] Another factor is that the pulsed operation of the capillary discharge allows it to operate at high pressures (100s to 1,000s atm), allowing efficient nozzle expansion even at the high temperatures ($T_p \sim 1eV$) required to achieve the desired specific impulses. The Advanced Concepts Group identified three tasks that were required to demonstrate proof-of-concept for capillary discharge based propulsion systems: demonstrate the required efficiency at desired specific impulses, demonstrate efficient and repeatable ignition, and demonstrate a robust propellant feed mechanism. The Advanced Concepts Group performed both experimental and computational efforts to address the first two tasks during the capillary discharge research.

The Advanced Concepts Group first built a simple performance model to gain a basic understanding of the operation and ultimate potential of capillary discharges. A 0-D time-dependent high-pressure slab capillary discharge model was created by Pekker.[10] It, for the first time, included a heat transfer radiation model based on a radiation database to self-consistently calculate the radiation heat flux at the thin transition boundary layer between the uniform plasma core and the ablative capillary walls. Dr. Pekker followed the progression of capillary modeling starting with the work done by Loeb and Kaplan[14], which used many assumptions to simplify a capillary discharge model. Later work done by Powell and Zielinski[15] added the Saha equation to more accurately calculate plasma composition and enthalpy. From these previous works, Pekker developed a model based on the conservation laws for mass and energy. The model also includes: a self-consistent heat radiation model; a model of the transition boundary layer to obtain the boundary conditions connecting the plasma core parameters with the parameters at the ablative surface; and the resistor-inductor-capacitor circuit model. Thus, the model allows the self-consistent calculation of plasma parameters and distribution of wall temperature verses time. More detailed information about the model and the computation results are given in the papers by Pekker.[10,11]

The model predicted the existence of two steady-state capillary discharge regimes: a high pressure regime and a moderate pressure regime. The two regimes converge at low plasma temperatures, and there is no steady-state solution for plasma temperatures lower than the convergence point. Both operating regimes may be attractive for thruster applications. The model also allows the calculation of the grey factor, the ratio of radiation flux at the transition boundary layer to the blackbody radiation flux with the same plasma temperature. The grey factor may vary from 0.02 to almost 1 depending on the plasma parameters and capillary geometry. Thus, the assumption that the grey factor is constant in a non-steady operation regime can lead to false results.

One simplifying assumption made in the 0-D model is that the radiation wall absorption coefficient is assumed to be constant which is not physically realistic for certain propellants such as polyethylene because the extinction coefficient of polyethylene depends considerably on wavelength and temperature, which may change significantly in the course of discharge. The model shows that a small wall material extinction coefficient leads to large energy losses from the capillary discharge (the heat is absorbed by the bulk of the capillary wall or just escapes the capillary), and to an initial spike in the plasma temperature. If the extinction coefficient is too small, the discharge may become extinguished because the temperature of ablative surface does not increase fast enough to compensate for the plasma exhaust from the open end of the capillary. Thus, for thruster applications the capillary wall material has to have a radiation absorption coefficient as large as possible to improve the energy efficiency of the thruster and to more easily achieve stable regimes of thruster operation. Results from the model also indicated that it was possible to achieve the desired performance levels which motivated experimental efforts to validate the performance estimates and evaluate potential ignition techniques.

A 1-D model was created by Sergey and Natasha Gimelshein in order to incorporate some of the flow field physics that were missing from the zero dimensional model.[12] Several noticeable discrepancies, in particular the prediction of a long extinction tail which was not observed experimentally, suggested that a more physically realistic 1-D model was required. The axially symmetric 1-D model was based on the work down by Edamitsu and Tahara[16]. Their computational and experimental work was on a smaller, lower energy, electrothermal PPT with energies of 5 to 15 J. While considerably different from the capillary discharge presented in this work, the same basic concepts and physics applied. Their unsteady numerical model simulated the electrical circuit, plasma flow, heat transfer to the wall, heat conduction inside the wall, and ablation. The assumptions used in the 1-D model are similar to the 0-D. The plasma flow was considered in ionization equilibrium, single ionized, one-fluid plasma flow, and in local thermodynamic equilibrium LTE. The effects of the magnetic field are not considered and the total pressure and electron number density were assumed to be radially constant. The results from the model are given below in comparisons with experimental results and in more detail in the listed references.[12]

The first step in the capillary discharge experimental research was to identify and quantify a reliable capillary discharge ignition technique. The traditional capillary discharge ignition technique, an exploding wire[17], was inadequate because it can lead to large variations in performance, contributes a significant fraction of the exhausted mass (not desired or included in the models), and it is fundamentally unsuited for spacecraft operations. Three different ignition techniques were evaluated: the traditional exploding wire ignition, Paschen breakdown ignition, and a three electrode ignition which are shown in Figures 20-22, respectively. In the exploding wire ignition technique a thin metal wire electrically connects the anode and cathode of the discharge. When a voltage is applied across the electrodes the thin wire shunts the current, rapidly heats, and then vaporizes to provide initial plasma that can conduct current and start the discharge. In the Paschen breakdown ignition, a residual gas is maintained in the capillary and the breakdown will occur whenever the potential applied across the electrodes exceeds the Paschen limit. This ignition technique is most relevant for multi-pulse operation when subsequent pulses may occur before all of the gas from the previous pulse has been expelled from the capillary. In the three electrode ignition system, a third electrode is added near the expelling end of the capillary discharge. A high potential is applied between it and the nearest electrode. A surface flashover occurs which provides the initial plasma necessary to start the discharge.

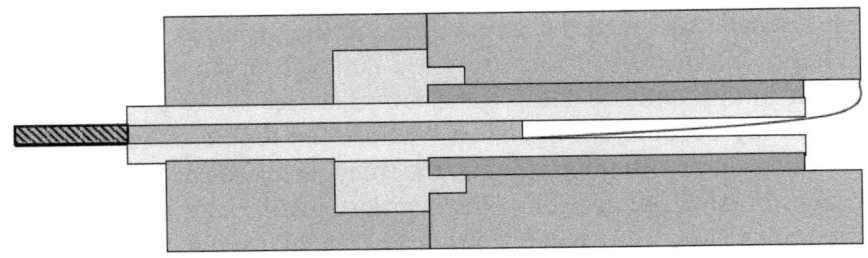

Figure 20. Capillary Discharge Configuration for Exploding Wire Ignition[12]

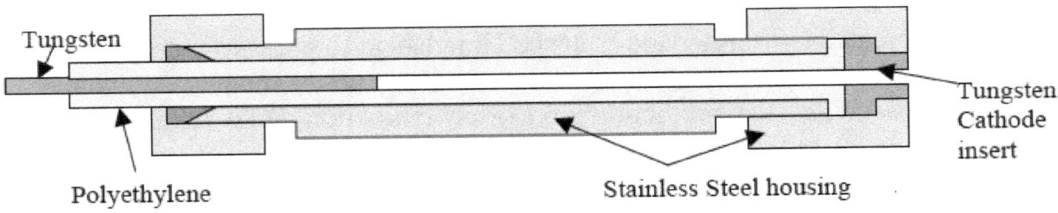

Figure 21. Capillary Discharge Configuration for Paschen Breakdown Ignition[12]

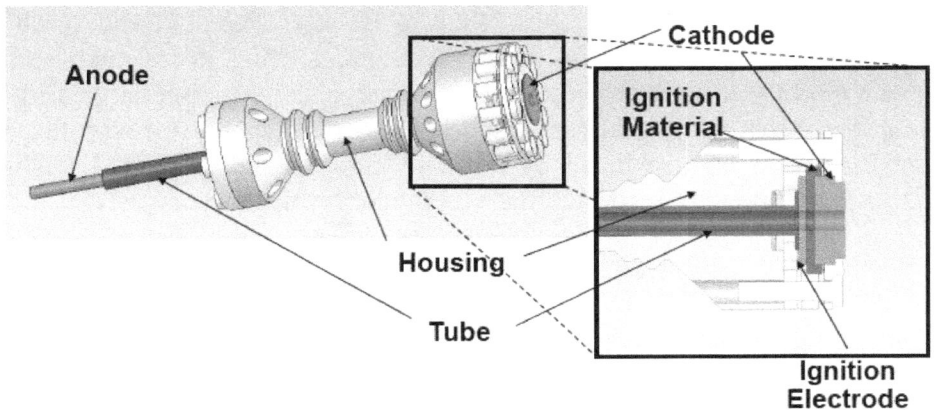

Figure 22. Capillary Discharge Configuration for Three Electrode Ignition[12]

Experimental testing of each configuration was conducted by placing them on a thrust stand in a vacuum chamber (as shown in Figure 23) and operating the discharges over a wide range of operating conditions. High voltage probes and Rogowski coils were used to measure the circuit conditions during the discharges. The masses of the capillaries were weighed before and after shots while the impulse of each shot was measured by the thrust stand. High speed video and emission spectroscopy were used to diagnose the plume for some of the tests. A complete summary of the experimental work is given in Dr. Pancotti's thesis.[12]

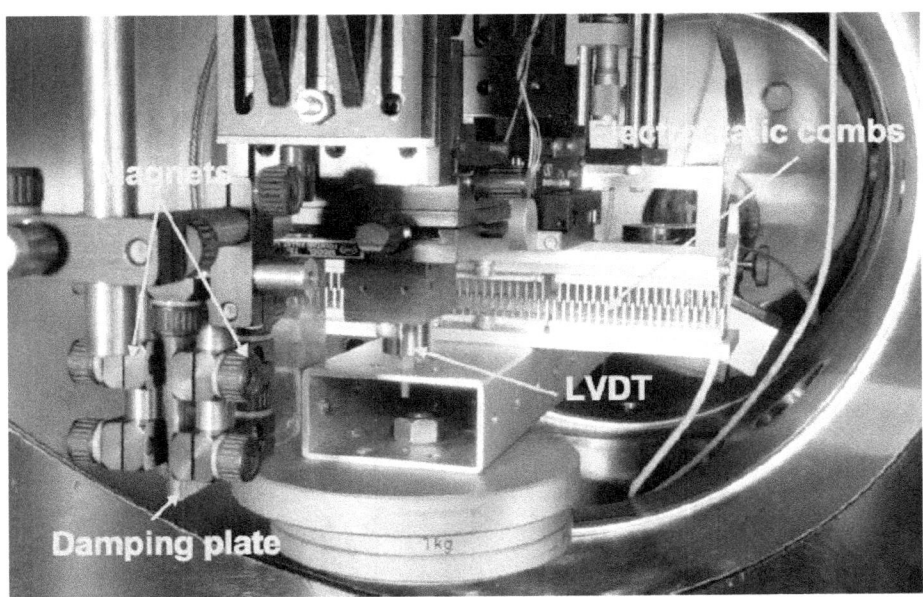

Figure 23. Thrust Stand With Calibration Combs in the Capillary Discharge Vacuum Chamber[12]

Only a limited selection of the results will be given here with the rest being contained in the references in the bibliography.[9-12] Figure 24 shows plots of the measured current during capillary discharge firings for the three identified ignition techniques. The current profiles are also compared to predicted profiles from both the 0-D and 1-D models. The different plots correspond to capillaries of different lengths (4 cm to 10 cm). There are measurable differences between the current profiles for different ignition techniques which is reasonable since they each ignite the plasma slightly differently and the discharges are never operating in a quasi-steady condition. It is shown that the models and experiments agree fairly well with the one exception being the long extinction tail predicted by the 0-D model. The predicted tail is particularly pronounced for longer tubes. As expected the 1-D model resolved the issue.

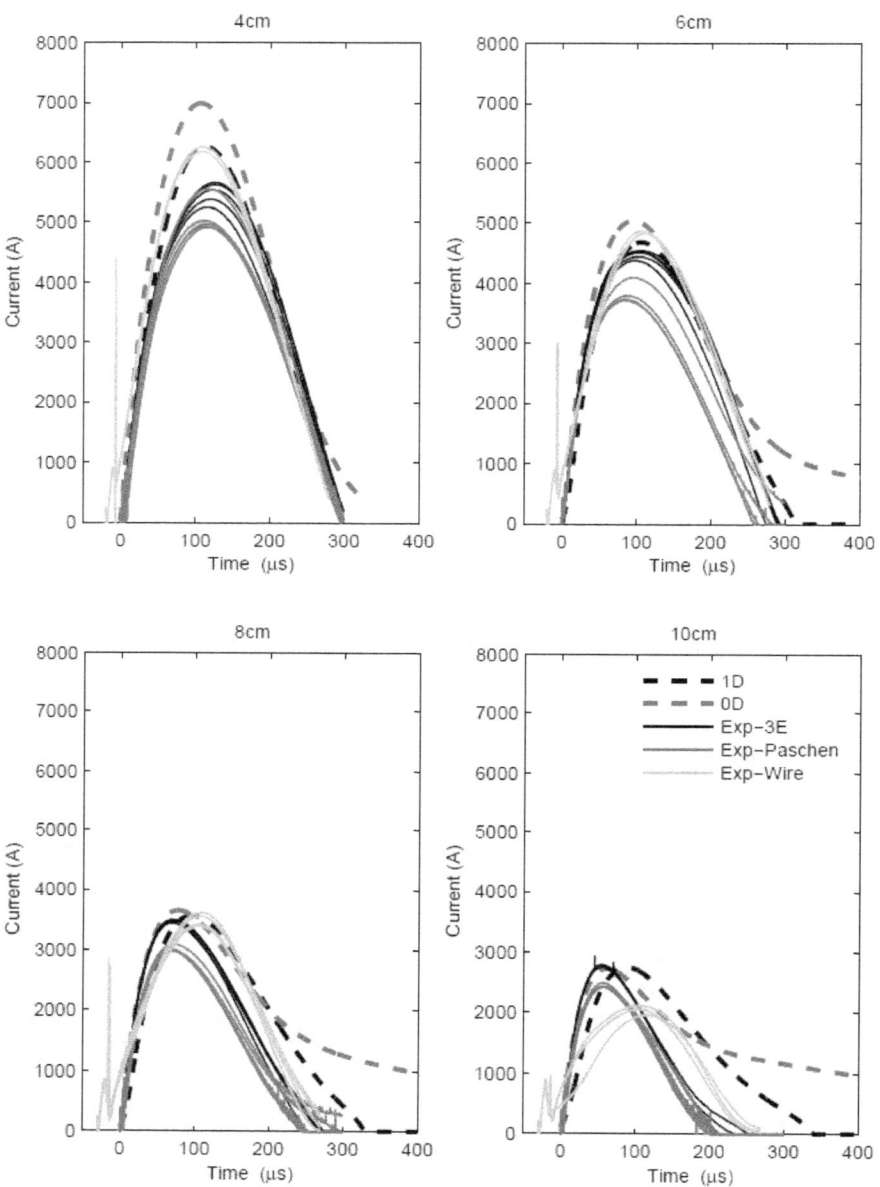

Figure 24. Discharge Currents for Three Ignition Methods using 4, 6, 8, and 10 cm Capillaries at 2500V[12]

Figure 25 shows the performance characteristics [(a) total impulse, (b) specific impulse, and (c) thrust efficiency] for capillary discharges operated at different pulse energies and using the three electrode ignition technique. It should be noted that these tests did not include a nozzle since they were focused on evaluating the ignition techniques.

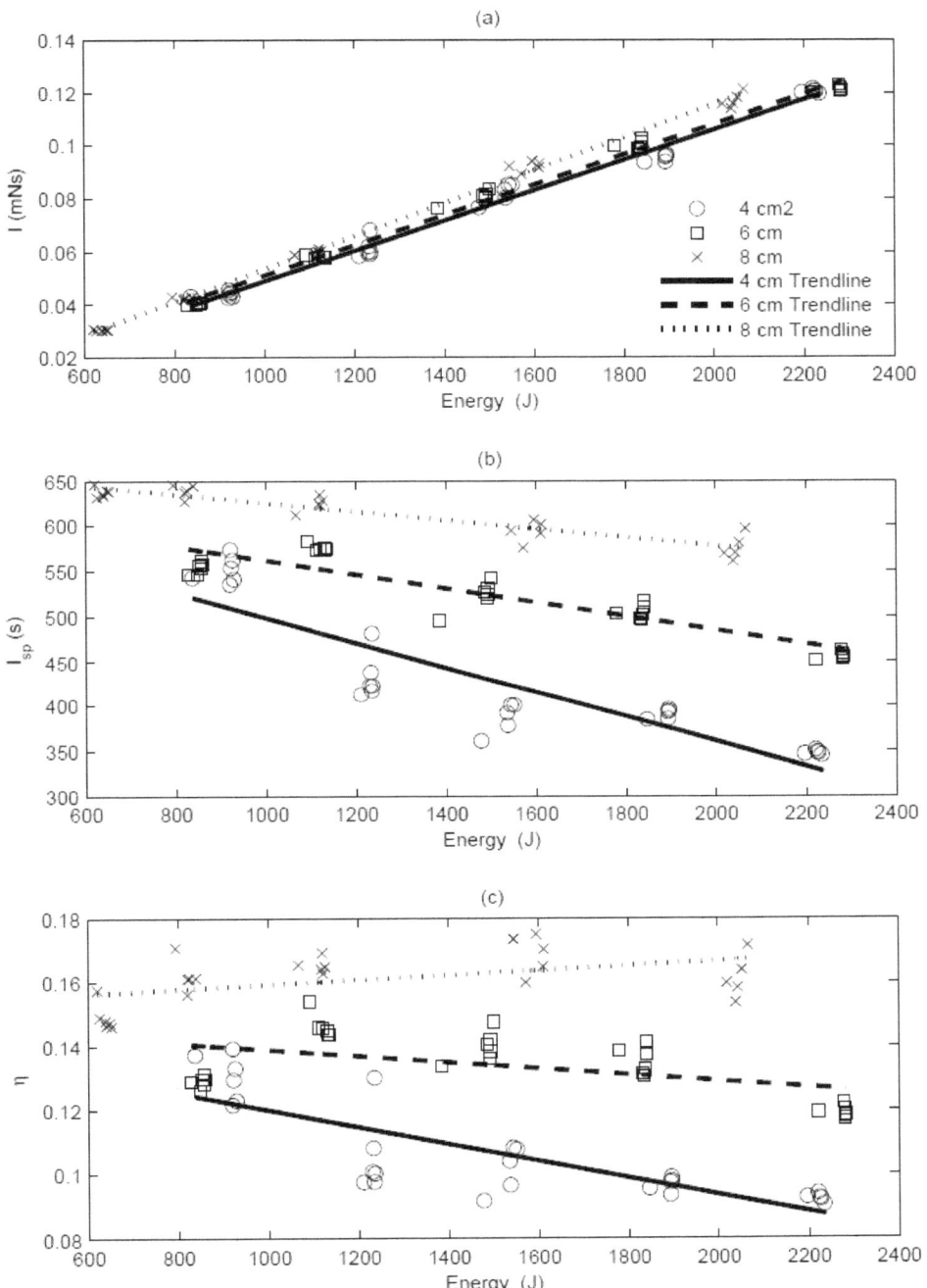

Figure 25. Impulse (a), Isp (b), and η_t (c) for 4, 6, and 8 cm Capillaries for Surface Flashover Ignition[12]

It was observed that the total impulse scales predominantly with pulse energy and the capillary geometry has only a minor effect on the total impulse. The specific impulse tends to decrease with increasing pulse energy as does the thrust efficiency which maximized at roughly 17% (for the case with no nozzle). Overall performance and circuit characteristics proved to be very consistent over the range of methods and large deviations were only seen during wire ignition testing. It is speculated that all ignition techniques, including the low current mode of the wire ignition, are in fact a Paschen-type breakdown in which conditions are created within the capillary that are ideal for electron cascade down the entire capillary length. In the high current wire ignition mode, it is believed that a more instantaneous and fully ionized plasma column is created by the chaotic nature of wire explosion which allows charge to move more rapidly and immediately. This could explain the higher current mode and its greater likelihood at shorter lengths. The tests conducted here have proven that three electrode ignition is reliable and repeatable without the need for manual replacement of an ignition wire or atmospheric control rendering it the best option currently available for adapting the capillary discharge to space propulsion applications. The developed three electrode ignition system has laid the groundwork for future capillary testing at the AFRL, including an investigation into the effects of nozzle expansion and alternative capillary materials, to constitute a full capillary discharge thruster.

To gain a better understanding of how the capillary discharge would perform as a thruster, a nozzle was added to the geometry, as shown in Figure 26, and a wide variety of potential propellant materials were tested, as shown in Table 6.

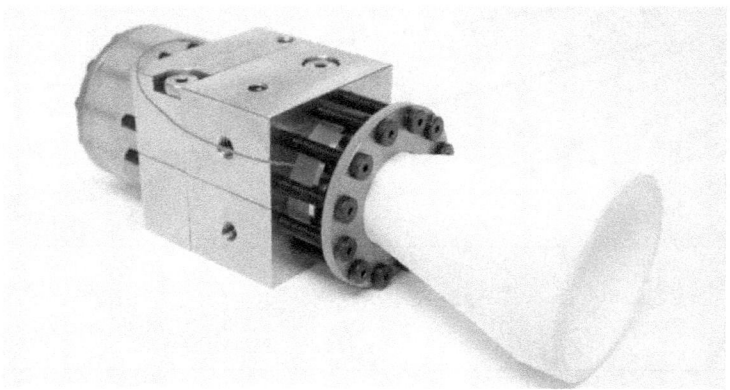

Figure 26. Final Capillary Discharge Configuration Shown with Ceramic Nozzle[12]

Table 6. Propellant Materials Tested in Capillary Discharge Research[18]

Tested Capillary Discharge Materials		
• HDPE	• PEEK	• FEP
• Nylon 6/6	• Pyropel HD	• PPS
• Molybdenum Disulfied Nylon	• Vespel	• Delrin
• Teflon	• K-Fel	• PTFE Delrin
• Graphite Teflon	• Rulon 123	• POM
• Fluorosint LF207	• Rulon 142	• Acetal Copolymer
• Fluorosint HPV	• Torlon	• Turcite
	• Radel	• PVFD

Pancotti gave a thorough discussion of the performance of the different capillary discharge materials.[18] Only the final conclusion will be given here. Shown in Figure 27 are grey lines that represent the performance curve fits for classes of flight qualified electric propulsion systems. It is clear that the region around 1000s is the only remaining region without the availability of flight qualified electric propulsion systems (a light grey wedge shows the area of interest). The performance of capillary discharges operating on high-density polyethylene HDPE (red), Teflon (blue), and Acetal (green) are also shown on the plot. The HDPE capillary discharge agrees very well with the performance predictions. The Teflon capillary discharge slightly underperforms the predictions. Acetal, however, significantly outperforms the predictions. It is theorized that this may be due to increased recombination in the nozzle leading to more energy recovery. Additional research would be required to fully determine the cause.

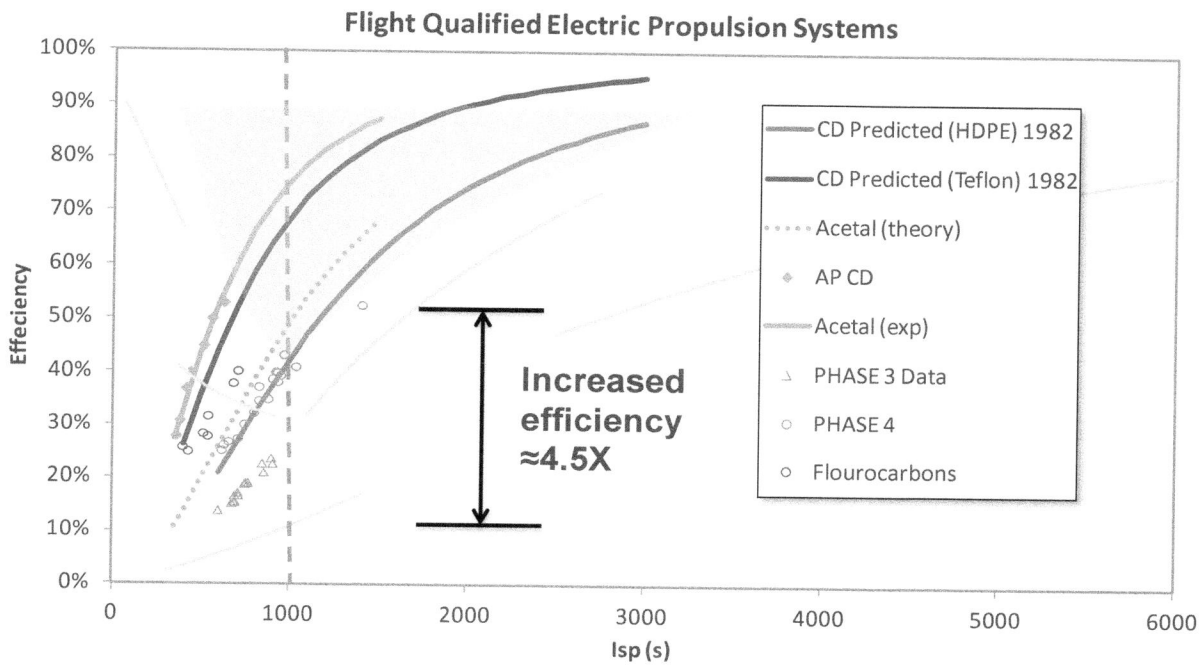

Figure 27. Capillary Discharge Performance Dependence on Material[12]

The capillary discharge research demonstrated a three electrode ignition system that performed well and was repeatable. A thorough computational and experimental campaign developed a general understanding of the capillary discharge physics. A series of tests evaluating different circuit parameters, discharge geometries and capillary materials concluded that it was highly likely that a thrust efficiency of up to 65% was achievable at a specific impulse of 1000s. The proof-of-concept for capillary discharge based propulsion systems could be completed by completing some additional materials tests and by developing a reliable propellant feed mechanism.

4.5 Macron Launched Propulsion[19, 20]

As discussed earlier, the Advanced Concepts Group determined that a propulsion system that could operate efficiently ($\eta_t > 50\%$) at intermediate specific impulses (Isp ~ 1000s) would yield a new capability for highly responsive satellites and would represent a practical first step in the development of an "ideal" thruster that could operate efficiently over a wide range of specific impulses (500s - 5,000s). Traditional electrostatic and electromagnetic propulsion systems have difficulty achieving high thrust efficiencies at these specific impulses due to the large energy cost required to produce the required ions. One strategy for overcoming these limitations is to use a pulsed inductive acceleration of a solid conducting shell or "macron" which would not require any ionization as suggested by Dr. Slough at MSNW.[19,20] Recent advances in energy storage and solid-state switching have enabled the use of peristaltic, pulsed inductive acceleration of non-ferritic particles for spacecraft propulsion. Macron Launched Propulsion (MLP) systems electromagnetically accelerate gram-sized aluminum particles (i.e., macrons) to achieve exit velocities between 5 and 10 km/s, achieving specific impulses between 600 and 1,000s. The macron propulsion system is similar to the FRC propulsion system discussed later except it expels a conducting shell of solid material instead of a plasma shell and, since its propellant is more massive, it naturally operates at lower specific impulses. A notional schematic of a 100 kW class MLP is shown in Figure 28 along with pictures of the 1 gram aluminum macrons used during the macron research, and the projected properties of a 100kW macron propulsion system. The power level of 100 kW would be the most convenient to build and test in the short term. There were two primary concerns that required addressing to show proof-of-concept for the MLP system: demonstrate that it can indeed achieve the required velocities (5 km/s – 10 km/s), and demonstrate that macrons ejected in space will not contribute to the growing space debris problem.

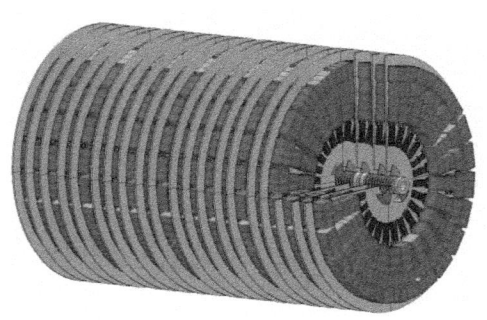 

Macron Parameter	Value
Macron Mass	*1 gram Al*
Coupling Parameter	*2.5*
Macron Velocity	*8 km/s*
Macron Kin. Energy	32 kJ
Per Pulse Energy	60 kJ
Rep. Rate @ 100 kW	2.5 Hz
Rep. Rate @ 400 kW	10 Hz
Total Thruster Length	2 m
Spec. Pow. @ 100 kW	0.4 kW/kg
Spec. Pow. @ 400 kW	1.6 kW/kg
T/P @ 800 s Isp	200 mN/kW

Figure 28. Projected 100kW Macron Propulsion System and Macron Geometries

Early estimates suggested that the MLP could be a highly efficient thruster with no ionization losses and may eventually achieve efficiencies upwards of 90% while operating at 600 to 1,000s of specific impulse (Isp).[19] The experimental portion of the project began by assembling a relatively simple 10-stage system to simply demonstrate the acceleration mechanism and achieve macron velocities of a few hundred meters per second. The 10 coil development test section is shown in Figure 29. Results of the velocity attained as a function of stage are shown in Figure 30 for two different capacitances. The time response of the higher capacitance stages was insufficient for optimal energy coupling which is why it did not achieve the same velocities in stage 5 and stage 6 as for the other capacitance. One of the most important lessons learned from the preliminary tests was that optimizing the design to efficiently couple energy at each stage was going to be the most difficult aspect of the research. Different macron materials (Al 6060, 2024, 7075) were also tested to determine their effect. It also showed a high repeatability, full cycle energy recovery, and performed > 500 shots with the same launch tube. The second phase of the project would have increased the exhaust velocity to over 1 km/s, but was not completed due to funding limitations.

Figure 29. 10 Coil MLP Development Test Section

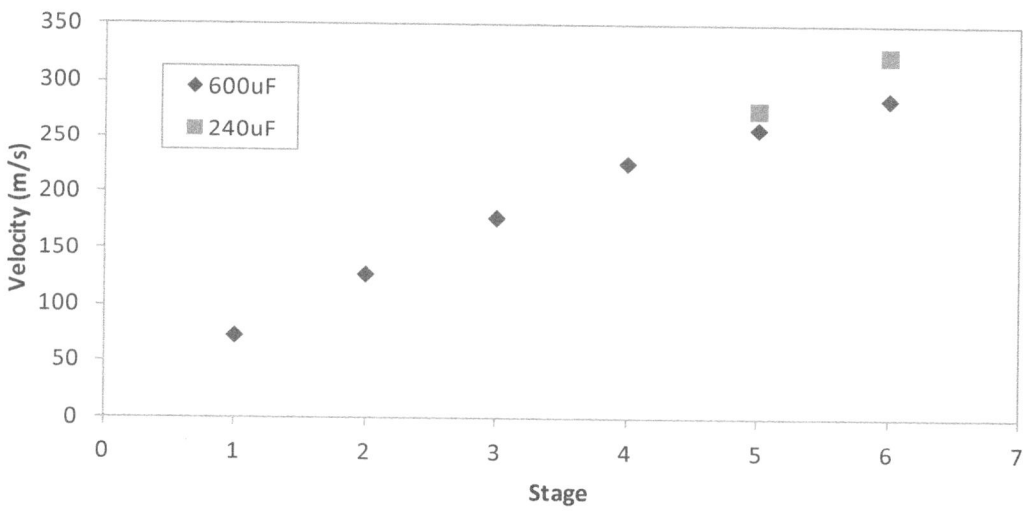

Figure 30. Achieved Macron Velocity for Six Stages

A parallel phase of the MLP research analyzed the potential effects of the ejected macrons on the orbital debris environment. It is critically important for this system to be utilized in a manner which minimizes its effects on the orbital debris environment. Satellite Tool Kit™ (STK) was used to determine the resulting trajectories of the fired macrons. Figure 31 shows the outcome of the analysis. For Low Earth Orbit (LEO) all exhaust velocity cases could lead to escape, orbit, or reenter conditions depending on the firing angle. Note that zero degrees correspond to firing the macron in the ram direction. For GEO on the other hand, just about all macrons escape with the exception of the 5 km/s exhaust velocities that are fired in the anti-ram direction.

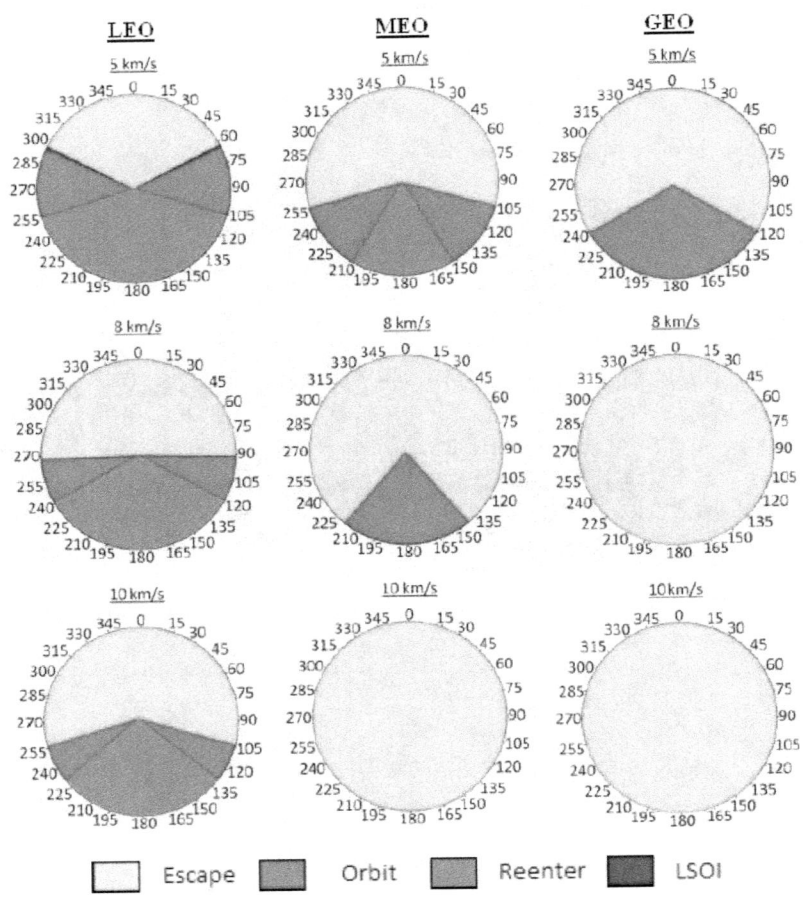

Figure 31. Macron Behavior for Various Orbit Conditions[20]

Trajectory analysis shows, in general, macrons which are fired with the majority of their velocity directed in either the RAM or anti-RAM directions, minimize the potential impact on the space environment. The three most important factors to consider before each firing are the macron's firing angle, exit velocity and the altitude at which the firing will take place. These three parameters combined dictate the level of a macron's impact. Before any orbital maneuvers are conducted, it is critical that spacecraft operators take precautions to ensure avoidance of orbital trajectories or collision course trajectories with neighboring spacecraft.

Orbital analysis supports the implementation of this system as an intra-LEO, LEO to GEO or GEO inclination changing propulsion system. Operating this system at GEO minimizes the potential impact of a macron due to the decreased energy levels required to place a macron on an escaping trajectory. In addition to GEO inclination changing maneuvers, this propulsion system could also be implemented as a GEO phasing or GEO to GEO rendezvous orbital maneuvering system. Additional details are available in the references.[19,20]

The MLP was also evaluated to address another challenge: *"Demonstrate a propulsion system that can allow satellite formation flying at significantly higher effective specific impulse (Isp_{eff} > 100,000s) and avoid contamination of other satellites in the constellation."* In addition to traditional orbital maneuvers, a single stage variation of a MLP system can be implemented as a momentum exchange device to enable satellite platoon formations which are impossible to sustain through the use of traditional propulsion techniques.[21] A sustained tandem satellite formation with a constant separation distance between satellites is not only achievable through the implementation of a single stage variation of a MLP system, but it can also significantly benefit the field of remote sensing by enabling bistatic synthetic aperture radar (SAR) operations and satellite platoon formations with a theoretically infinite operational lifetime. Figure 32 illustrates the suggested concept where a continuous stream of macrons are maintained between two satellites in a formation by constantly collecting and relaunching solid macrons between the satellites. In Figure 33, the estimated maximum miss distance as a function of firing velocity for different separation distances is given. Collectors on the order of 5 cm should be sufficient for this application.

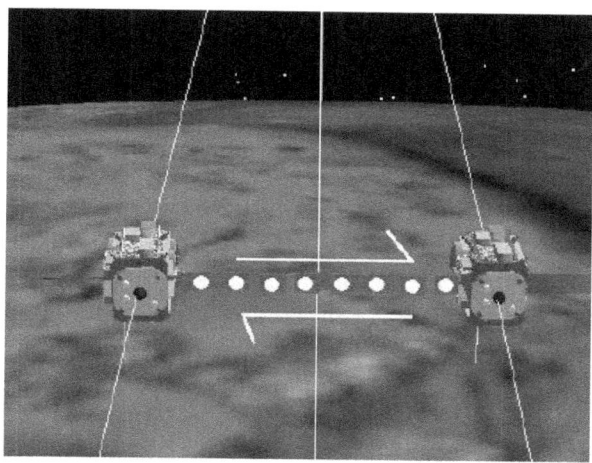

Figure 32. Momentum Exchange Concept for Formation Flying Satellites[21]

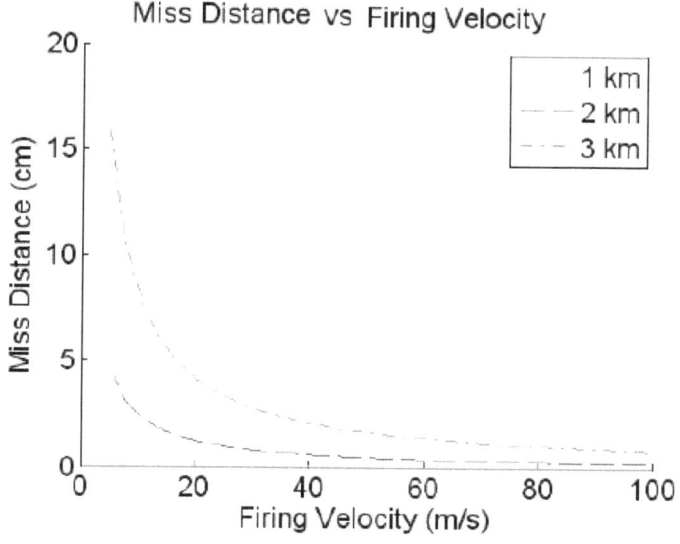

Figure 33. Maximum Macron Miss Distance[21]

The trajectory analysis provided by a miss distance code supports the implementation of macron propulsion technology as a momentum exchange device for satellite platoon formations. Furthermore, the thrust and optimal system configuration requirements to sustain a tandem satellite formation are well within the operational performance range of a single stage variation of a MLP system. Furthermore, the operational lifetime of a tandem satellite formation utilizing a single stage variation of a MLP system as a means of momentum exchange is theoretically infinite. Additional development work would be required to demonstrate the proof-of-concept of the collection system, but the single stage launch system has been proven. Additional information is contained in the reference by Schonig.[21]

4.6 Field Reversed Configuration (FRC) Propulsion

The FRC propulsion was initiated early in the Advanced Concepts program, after discovery of a previous work done by J. Slough, though the small business MSNW Inc. and the University of Washington, funded by the NASA Institute of Advanced Concepts (NIAC).[22] In that concept, dubbed the Propagating Magnetic Wave Plasma Accelerator (PMWAC – notwithstanding the omission of one letter to construct the acronym) and shown in Figure 34, a stable plasmoid is formed (the FRC) and accelerated along a channel as a result of the action of pulsed coils which sequentially exert a magnetic pressure gradient to accelerate the plasmoid. In essence, the device operates like a coil-gun, where the plasma "bullet" rides on a wave of magnetic pressure that is externally controlled. Here, the need for a mechanical target is eliminated by using high-conductivity plasma, which can be rapidly injected in the chamber. Given the low mass of the plasmoid, the projected specific impulse of the device was very high, from 10,000 to 100,000 seconds or more for hydrogen propellant, making it ideal for high-power, deep-space missions.

Figure 34. Prototype PMWAC Built During NIAC Phase I Effort (Length 2 m)[22]

Such mission and performance figures are of lesser interest to the USAF, but the concept appeared promising, provided it could be rescaled and tuned towards higher thrust, i.e., higher plasma density, at the expense of the Isp for constant power. Thus, it would be useful to switch to higher mass propellant, increase the gas density and shorten the acceleration tube. Dr. Cambier decided then to provide funding to MSNW Inc. to further develop the idea. The NIAC project did not move into a Phase II, but we believed that we could leverage the hardware already built. MSNW instead suggested that a much better device could be built from scratch, using more modern and COTS technology for the power processing; the Magnetically Accelerated Plasmoid (MAP) project was started as such, with the following guidance from

AFRL: a) heavy propellant; b) maximum plasma density; c) shorter acceleration; and d) magnetic energy recovery. The MAP experiment with 8 accelerating coils is shown in Figure 35. The last item aimed at maximizing the efficiency of the device; it was clear from the previous studies and discussions with J. Slough that the pulsing of the coils for plasmoid acceleration was only able to transfer a fraction of the electrical energy in the coil currents to kinetic energy of the plasma. This is due to the short interaction time and exponential decay of the coupling between the coil and the FRC, a problem already observed in another thruster concept, the Pulsed Inductive Thruster (PIT).[23] The current left in the coil circuit after the acceleration occurs is subject to a dampened oscillation ("ringing"); this energy is thereby transformed into heat. Instead, it is possible to design the circuit to recover most of this energy, i.e., by charging a capacitor. The estimates of the overall system efficiency[1] when this energy recovery is included were of the order of 80-85%, a significant benefit, especially for high-power EP systems, where heat rejection becomes a major issue. Thus, the MAP concept appeared very well-suited for high-power EP, one of the key objectives of the Advanced Concepts program at that time (with applications such as Orbital Transfer Vehicle – OTV).

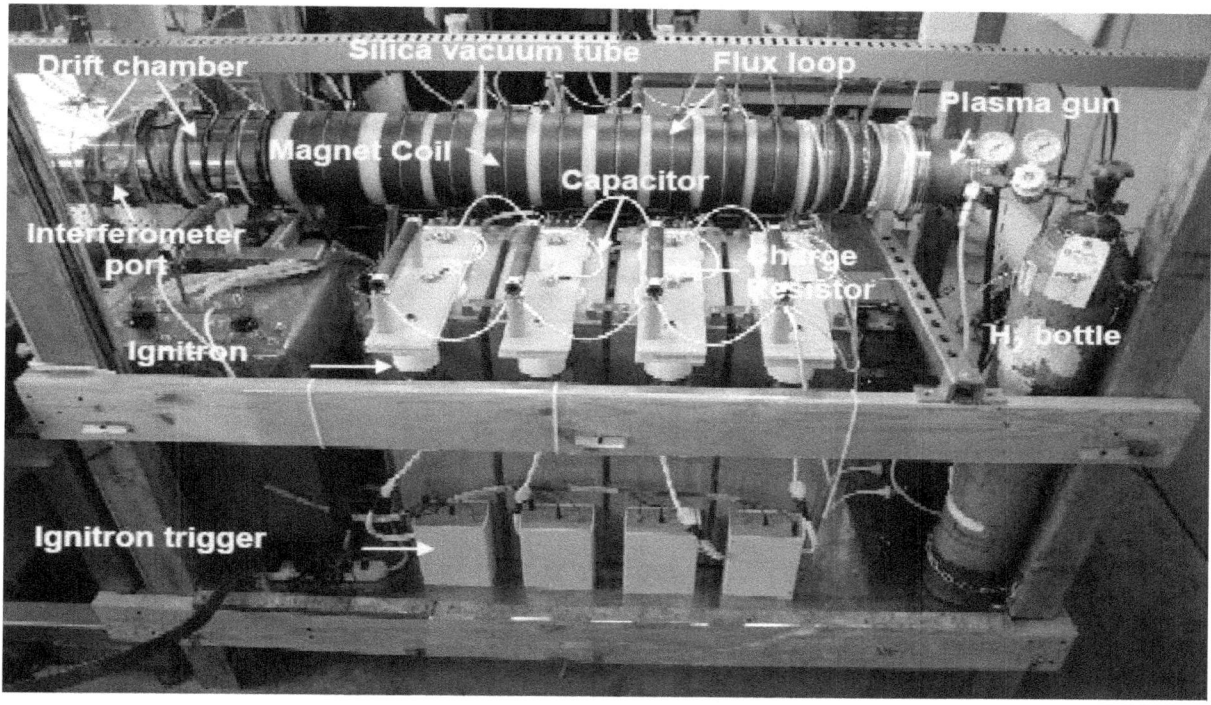

Figure 35. MAP Experiment with Eight Accelerating Coils[24]

The MAP program was quite successful; a new device was built and tested and plasmoids were successfully accelerated in the device to high velocities. Figure 35 shows the device built and Figure 36 is an example of the trace of flux loops along the tube, showing the passage of the FRC. A visible light Mach-Zender interferometer was used to measure on-axis plasma density;

---

[1] MSNW was pressed on these figures and we occasionally corrected the estimation procedure, but unfortunately, the MAP program never provided a definitive experimentally measured value of this energy recovery efficiency. However, the basic approach and efficiency gain has been confirmed by follow-on programs with a single acceleration stage.

most of the studies were performed with D_2 gas with 10% He added for spectroscopic determination of ion temperature.

The MAP program also investigated the possibility of multiple FRC acceleration, by using the ringing in the coils to induce a plasmoid formation at each half-time step of the oscillating circuit; this approach was only partially successful, as only FRCs separated by a full period of the circuit were stable; the others were propagating into a bias field with the wrong polarity and were quickly dissipated.[24] This observation correlated with two ideas which developed soon after the project initiation; a) efficient FRC formation and b) mass entrainment. The standard approach to FRC formation is a theta-pinch; starting from a gas injection into a tube with an ambient ("bias") longitudinal magnetic field, coils are pulsed which induce a magnetic field in the opposite direction to the bias field. The rapid change in magnetic field creates an ambient electric field inside the tube. If the gas is sufficiently pre-ionized, this induced field accelerates electrons and leads to a rapid growth of the ionization fraction and conductivity until a fully ionized plasma is formed. The high conductivity traps the bias field inside the plasma while the external magnetic pressure from the pulsed coils further compresses and heats the plasma, while replacing the bias field on either side of the plasma by a stronger field of opposite polarity. The sequence of events is shown in Figure 37. The end-points of the plasma on the axis are reconnection points for the magnetic field. This leads to the typical FRC structure with closed poloidal field lines, a high-β configuration.

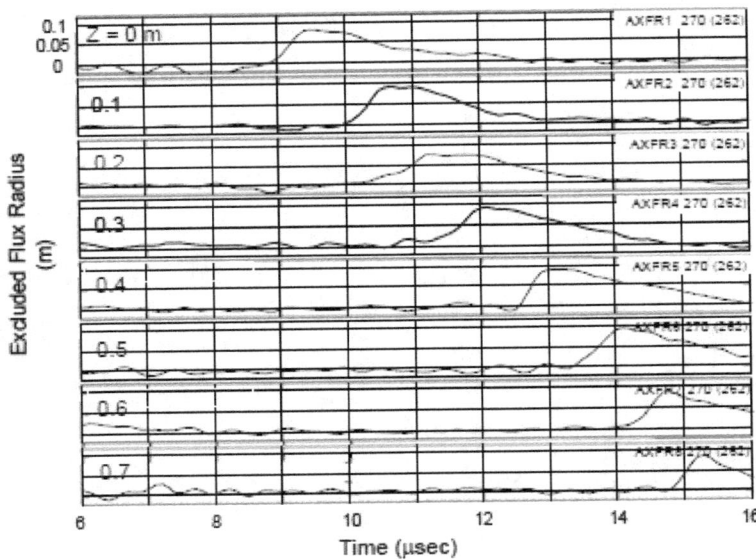

Figure 36. Trace of Excluded Magnetic Field, Indicating the Presence, and Transport of FRC[24]

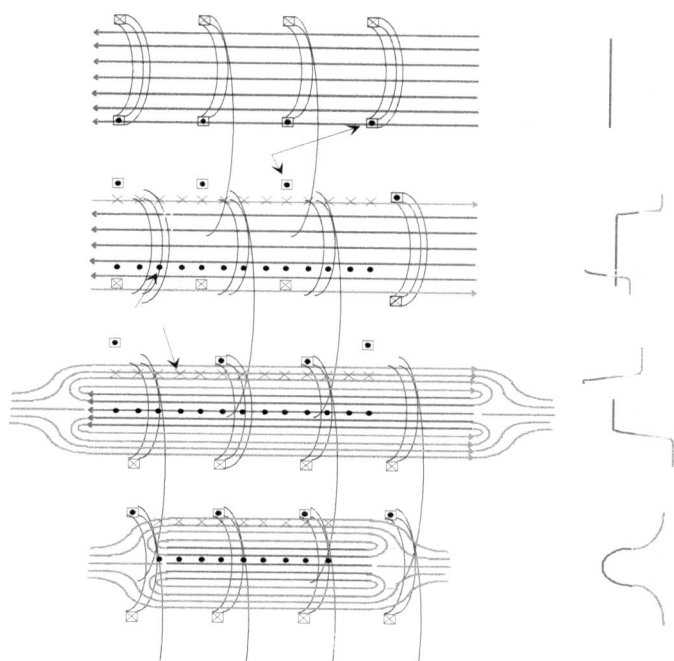

Figure 37. Typical Sequence for FRC Formation by Theta-Pinch.

However, this approach requires a very rapid rise of current in the coil circuit, in order to generate an electric field high-enough for rapid ionization. This creates severe constraints on the power processing unit (PPU) and either limits the power and size and/or increases the PPU weight, cost and failure risk. Alternatives to this approach have been investigated; one of them is the "slow-source", used in an annular configuration by researchers at the Space Propulsion branch of AFRL-Edwards (AFRL/RQRS). The other is the Rotating Magnetic Field (RMF) configuration which Slough was developing at the University of Washington.[25] The RMF approach allowed the use of COTS devices for the PPU, solid-state, i.e., light-weight and very reliable. Furthermore, decoupling the FRC formation from its acceleration allows much greater control of the repetition rate and eliminates the need for a bias field. Thus, it was determined that the RMF, combined with rapid expulsion of the formed FRC and magnetic energy recovery would be the next step in the development of this thruster concept. Subsequent development was accomplished through the combination of several programs from AFOSR (6.1) and AFRL (6.2), as well as SBIR efforts, and was administered by AFRL/RQRS.

The second observation concerned the interaction with ambient gas. Since a significant performance factor is the cost of ionization, it is desirable to minimize the amount of plasma necessary for coupling to the magnetic field (providing the acceleration force) while maximizing the mass for thrust enhancement. This led Dr. Cambier to suggest the concept of neutral gas entrainment, as an extension of a magnetic "piston" concept that was initially suggested in the context of magnetohydrodynamic enhancement of pulsed combustion systems. In this case, the plasma (FRC) is created and accelerated while pushing on an ambient, neutral gas. If residual gas in the FRC thruster is the working fluid, this provides the means of reducing the injection losses. However, the idea was suggested also as a means to operate the FRC thruster as an air-breathing EP system in hypersonic, high-altitude flight. The device would essentially operate like an ejector, increasing thrust by operating on a by-pass flow. However, because the FRC could easily be accelerated to high velocities, it is not necessary to slow down the incoming air flow to increase its pressure or the coupling to the primary (high specific energy) stream; thus the concept would not suffer from dramatic drag losses from an inlet or excessive heat loads from nearly stagnated flow[2].

The concept of neutral entrainment was further analyzed and thrust enhancement efficiency was defined after iterations between MSNW and AFRL. Experimental and computational investigations of the concept are currently under-way through AFOSR funding.

Overall, the FRC propulsion project was highly successful and is a shining example of technology transfer by the Advanced Concepts program; the next-generation designs of the FRC thruster have been tested in ground facilities, several Air Force and DOD programs have provided subsequent funding to further mature the technology, including NASA as well. The device is now much more compact, fine-tuned for AF applications and better characterized; its performance characteristics make it a significant advancement in space propulsion technology and provide the US with a clear technological advantage over world-wide competitors. Developments towards more advanced versions are still under-way, with the promise of yet even higher performance and a wider range of applications.

4.7 Pyroelectric Thruster

In the pyroelectric phenomenon, a temperature change applied to a select type of crystal can produce very high voltages across the crystal (~100 kV) due to small changes in the positions of atoms within the crystalline structure. Cubesats or other small satellites requiring very simple power/propulsion systems could employ the effect to produce the potential difference required for electrospray (colloid or FEEP) propulsion or perhaps even other electric microthrusters. Such systems would not require solar panels or batteries to power the propulsion system and would, instead, power it from the changing temperatures experienced due to the transit of the spacecraft through illuminated and shaded portions of its orbit. The goal of the pyroelectric thruster project was to determine the overall performance potential of such a concept and determine how well it could address the challenges identified for in-space propulsion.

The pyroelectric thruster project aimed to explore the parameter space of an electrostatic thruster with a field emission ion source driven by a pyroelectric crystal. Figure 38 shows an arrangement that was built at UCLA to generate an ion current via field ionization of deuterium gas. When used as a thruster the crystal will be coated with a liquid metal such as Cesium. The electric field which develops on the tip due to heating of the crystal will drive a Cesium ion current off into free space. The Cesium exhaust is then replaced by an inflow of liquid over the

---

[2] This observation applies to the engine itself, and is valid as long as the rest of the system does not provide a high frontal area where this drag and heating can occur; this implies, of course, a much more slender shape than the typical satellite.

crystal to the shank. The left panel of Figure 38 shows a lithium tantalate crystal with an electrode and a tip. The right panel shows the electric field near the tip and the trajectories that the emitted particles follow. Heating the crystal by tens of degrees leads to potentials on the order or 100 kV. The thruster can be analyzed in two limits: in one limit a single tip is used. The other limit, the saturation limit, employs multiple tips so that the limiting current is due to the pyroelectric current of the heated crystal.

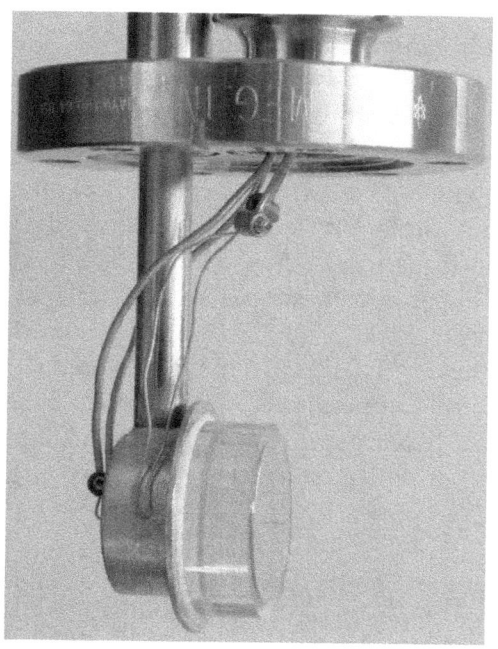
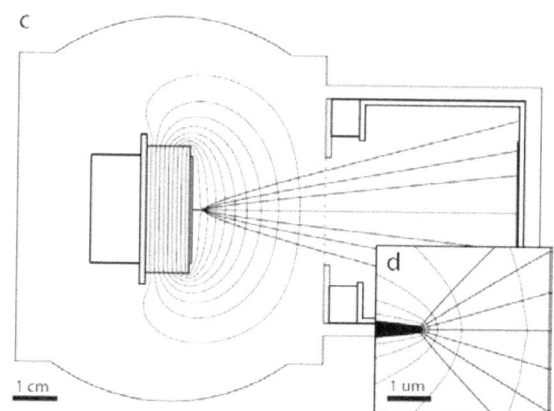

Figure 38. Crystal with Field Ionization Tip (Left), Equipotentials and Ion Trajectories.

It is assumed that the pyro-electrostatic thruster can be reduced to the following components: a source of metallic ions for field evaporation, an electrostatic accelerator based on a single face of a pyroelectric crystal with a tip mounted thereto, and in the case of multiple tips a grid electrode[26]. A key assumption to be made is that the electrode and thrust can be modeled in terms of a one dimensional flow[27]. Furthermore it is assumed that the field near the tip serves only to evaporate and ionize the atoms form the surface, and not to accelerate them. Once an ion current has been generated by the field emission tip, its role in determining the final exhaust velocity has ended. The exhaust velocity is determined by the surface potential of the crystal.

A simple analysis was performed to estimate the possible performance of a pyroelectric thruster. Equation 8 gives the potential difference, V, that can be produced across a pyroelectric crystal.

$$V = G \frac{p(\Delta T)H}{\kappa \varepsilon_0} \qquad (8)$$

Where ΔT the temperature change relative to T_0, the temperature of a reference state which for the purpose of this analysis we will be room temperature. H is the thickness of the crystal, D is the crystal diameter, k is the dielectric constant, and p is the pyroelectric coefficient with units of

Coulomb/m^{2}K, $1/4\pi\varepsilon_0 = 9$ x 10^9Jm/C^2, and G is a dimensionless quantity of order unity which includes corrections to (8) arising from the geometry of the crystal and the dependence of 'p' and κ on temperature. For the purpose of analysis we will take G=1. Results obtained with (8) will then be reliable to within factors of 2. If greater accuracy is desired then it will be necessary to include the variations mentioned above.

As an example consider lithium tantalate for which: p = 19nC/cm^{2}K and κ~40 [see reference 28]: taking H = 1 cm (D = 3 cm), one verifies that a temperature change of 25 K yields a potential of 130,000V. Including effects due to, finite geometry and the temperature dependence of p would give a result that is about 25% smaller. For lithium niobate the numbers are about the same. Other crystals such as Triglycine Sulfate (TGS) have a larger p but a smaller accessible range for ΔT. Crystals such as Strontium Barium Niobate have a pyroelectric coefficient that is over ten times bigger than for lithium tantalate, but these crystals also have a dielectric coefficient that is over ten times larger. So according to Equation 8, they will not make superior pyroelectrostatic accelerators. Whether the accessible parameter space of materials science includes crystal accelerators superior to lithium tantalate /niobate is not yet determined experimentally. According to tables of material parameters Thallium Arsenic Selenide (TAS) may be worthy of consideration. Finally the range of temperatures over which V can be increased by varying temperature is very large for lithium tantalate. Larger changes in temperature can easily access 200,000V potentials, provided that flash over is properly suppressed.

The exhaust velocity, v, is determined by balancing the kinetic energy by the accelerating potential[29]. Using the definitions:

e = electric charge = 1.6 x 10^{-19}Coulombs

n = degree of ionization

M = mass of the propellant liquid metal ion.

One finds:

$$v = [2enV / M]^{1/2} \tag{9}$$

As an example consider:

Cesium for which M_{Cs}=133 amu = 2.2 x 10^{-25}kg. For a potential of 100KV one finds:

$$v = 3.8x10^5 m/s \quad \text{for n} = 1$$

$$v = 5.4x10^5 m/s \quad \text{for n} = 2$$

The dependence $v(V)$ for n = 1 is shown in Figure 39.

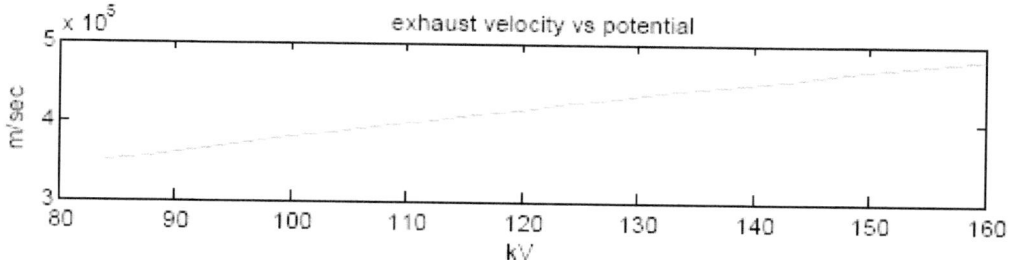

Figure 39. Typical Dependence of Exhaust Velocity on Accelerating Potential

The current, I, of ions off of the tip will be discussed in two limiting cases. For the case of one tip the current is limited by the emission of micro/nano droplets from the liquid that coats the tip. In this case the maximum current is:

$$I_1 \sim 10^{-4} \text{ Amps for one tip.}^{30} \tag{10}$$

This value is considerably lower than limits imposed by effects due to space charge.[27] When many tips or a blade are used the maximum current may in some cases be limited by the pyroelectric charge that can be generated on the surface of the crystal due to the applied thermal cycling. For this case we set the cycle time, Δt, and the current density, j [A/m²]. Using the definition of 'p' one finds:

$$j = p \frac{\Delta T}{\Delta t} \tag{11}$$

so that:

$$I_p = \frac{\pi D^2}{4} \frac{p \Delta T}{\Delta t} \tag{12}$$

A third case was also considered in which the current was consistent with experimental values expected to be obtained at UCLA with straightforward modifications of a previous apparatus[31]. This smaller current is:

$$I_0 \sim 100\text{nA.} \tag{13}$$

For a thick crystal the heating rate is sufficiently slow that (11) is not an improvement over (13). If, however, the active element is made from laminated crystals that are 1/10th the total thickness with heaters in between the crystal layers then Δt can be reduced to say 1 second. In this case a ΔT of 25 K yields from (11) a current density of 500nA/cm² which for a 3 cm diameter crystal is 3.4 µA, about 35 times higher than the value (13) which is expected to be achieved.

The thrust, F, is due to the reaction force of the exhaust and is given by:

$$F = \frac{I}{e} \sqrt{2eVM/n} \qquad (14)$$

and the thrust density, f, is given by

$$f = F/(\pi D^2/4) \qquad (15)$$

For the case that will soon be experimentally accessible, i.e., a single tip with 100nA current, we find:

$$F_0 = 50nN \qquad (16)$$

for, n=1, and an accelerating voltage of 100KV. The thrust density for the case where the pyroelectric current limits the ion thrusting current is:

$$f = \frac{p}{ne} \frac{\Delta T}{\Delta t} Mv \qquad (17)$$

For a temperature change of 25 K taking place over 1 second, the thrust density is

$$f_p = 275nN/cm^2 \qquad (18)$$

The heat energy supplied in one thermal cycle is:

$$Q = \pi D^2 HC\Delta T/4 \qquad (19)$$

Where C is the heat capacity ($\sim 3.J/cm^3 K$) for lithium niobate.

Another example of an interesting crystal is Tl_3AsSe_3 for which p = 350nC/cm^{2}K [reference 28]; and k ~ 35.

Although the UCLA group had no experience with this crystal it was known for its application as a thermal sensor. As an example of a different system the calculations based upon a crystal with the properties of this compound are presented in Table 7. It was assumed that the pyroelectric coefficient of TAS can be maintained over a range of tens of degrees. It was shown that the high pyroelectric coefficient of this crystal allows for a large current so that in a multi-tip arrangement a temperature change of 25 K for a 1 mm thick sample can generate 5 μN of thrust.

Table 7. Expected Pyroelectric Thruster Performance

			Accelerating Potential kV	Exhaust Velocity 10^4 m/s	Current μA	Thrust nN	Thrust Density nN/cm^2	Moles for Impulse of 10,000 Ns
LiTaO$_3$*								
	ΔT=25K	1 tip	130.	3.8	.1	50.	7.	1.
	ΔT=25K	saturated, Δt=1s	130.	3.8	3.4	2,000.	275.	1.
	ΔT=40K	1 tip	200.	4.7	.1	60.	8.	1.
	ΔT=40K	saturated, Δt=1s	200.	4.7	5.4	4,000.	550.	1.
Tl$_3$AsSe$_3$**								
	ΔT=25K	1 tip	260.	5.3	.1	70.	1,000.	1.
	ΔT=25K	saturated, Δt=.1s	260.	5.3	6.8	5,000.	70,000.	1.
	ΔT=40K	1 tip	400.	6.5	.1	84.	1,200.	1.
	ΔT=40K	saturated, Δt=.1s	400.	6.5	11.	10,000.	140,000.	1.

* 1cm thick x 3cm diameter

**1mm thick x 3mm diameter; entries based upon tabulated material constants

n=1, Cesium propellant

The tip geometry and voltage must be chosen so that the field at the tip is strong enough to lead to field ionization or field desorption of the liquid metal which coats the shaft. Typical fields are in the range of 10's of volts per nanometer. High fields are initiated not only by the radius of curvature of the tip but also by the Taylor cone which is formed by the liquid metal in the presence of the field[32,33]. When multiple tips are used a grid must be in place which grounds the field between the tips[26]. The electrical conductivity of the crystal must be sufficiently small so that during the time of cycling it does not short out the pyroelectric charging.

The performance criterion in Table 7 of 10,000Ns of integrated impulse is a specification for the Laser Interferometer Space Antenna LISA project[34]. The device is very compact and light-weight: the accelerator (crystal) is in itself the capacitor, and there is no need for additional circuitry to generate the high voltages. The device operates by transforming low-grade energy (heat) into kinetic energy, i.e., does not suffer from the conversion inefficiencies that typically plague electric propulsion systems, and could simply rely on passive solar heating.

The initial analysis indicated that it was possible to create usable amounts of thrust with the pyroelectric thruster necessitating experimental demonstration of the predicted performance levels. The research group at UCLA then designed and built a second generation ion source shown in Figure 40. The various improvements relate to the geometry so as to minimize, flash-over or simply sparking. Key to the performance of the thruster is a new method that was developed in order to remove imperfections from the tip where field ionization takes place. This is displayed in Figure 41.

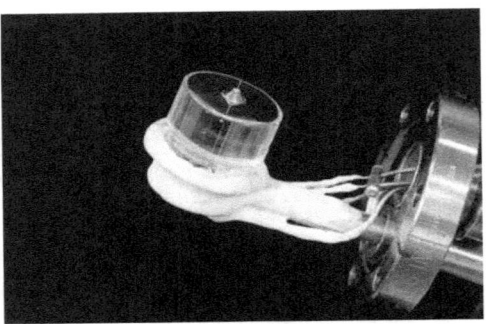

Figure 40. Second Generation Pyroelectric Thruster

Figure 41. Tungsten Tip out of the Box (Left); Tip After In-House Annealing Process (Middle); Tip After Explosion Due to Large Ion Current (Right)

Use of the in-house annealed tungsten tip led to dramatically improved stability of the field emission ion current that could be achieved with the 3 cm diameter crystal shown in Figure 40. The most stable controllable ion current to date is shown in Figure 42. Here the current is made up of deuterium ions so that their properties could be verified by nuclear fusion at a target [a rate of 500,000/sec has been achieved, unpublished]. The energy of the ion is about 100keV.

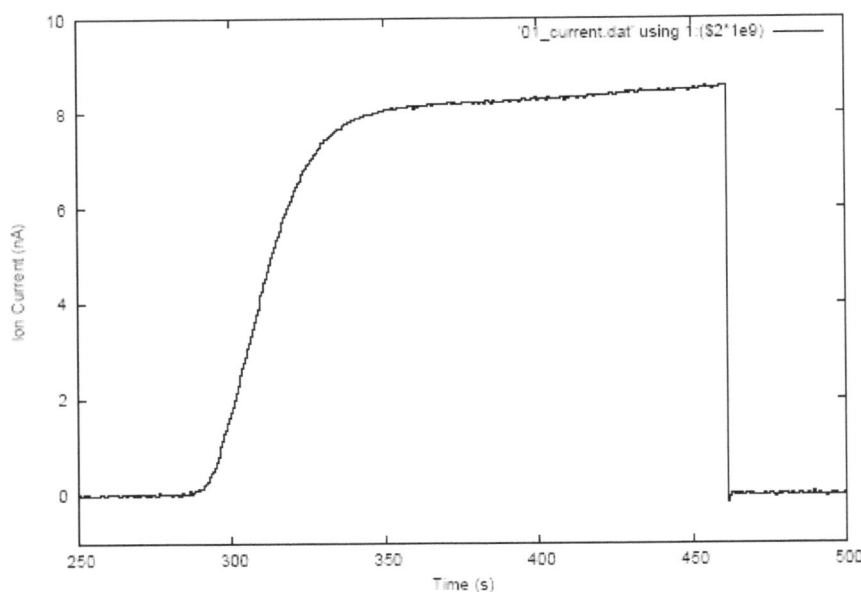

Figure 42. Ion Current of D from a Heated LiTaO3 Crystal With in-house Annealed Tip is Stable for 2.5 Minutes

The reaction force on the crystal which generates the potential for field ionization is: 1.nN for D nuclei and 10nN for heavy nuclei. Currents that are higher by a factor of 2 have been observed. The main advance of this effort has been to achieve controllable response that can be held for a time scale of minutes. A large percentage of this effort went into controlling flashover. In the future it is recommended that multiple tip assemblies and pixilated transducers assemblies be constructed.

4.8 Liquid Droplet Thruster[35]

As discussed earlier, it would be advantageous to "demonstrate a propulsion system that can allow satellite formation flying at significantly higher effective specific impulse ($Isp_{eff} >$ 100,000s) and avoid contamination of other satellites in the constellation." Dr. Ketsdever suggested investigating the potential of maintaining a continuous stream of liquid droplets between two satellites to address this challenge.[35] The liquid droplet thruster research effort evaluated momentum exchange through fluid streams as a means of maintaining side-by-side spacing between a pair of formation flying satellites. Droplet streams of very low vapor pressure silicone oil (such as Dow Corning 705) are generated on each spacecraft and projected through space to a receiving satellite. The receiving satellite collects the droplet stream and pumps the fluid to the droplet generator where a return stream is produced and sent back to the originating satellite. Therefore, tandem satellites could be envisioned as using streams of small silicon oil droplets continuously exchanged between them to produce the force required to maintain constant separation as seen in Figure 43. As Figure 44 shows, the thrust expected from the liquid droplet thrusters would be sufficient to maintain kilometer scale separation with small satellites. An additional side benefit of such a system is that it could also benefit the thermal control system by using the propulsion system itself as a liquid droplet radiator. Liquid droplet thrusters have several advantages over similar solid particle concepts: they have easier collection and transporting within the spacecraft processes and they have a higher useful propellant storage density.

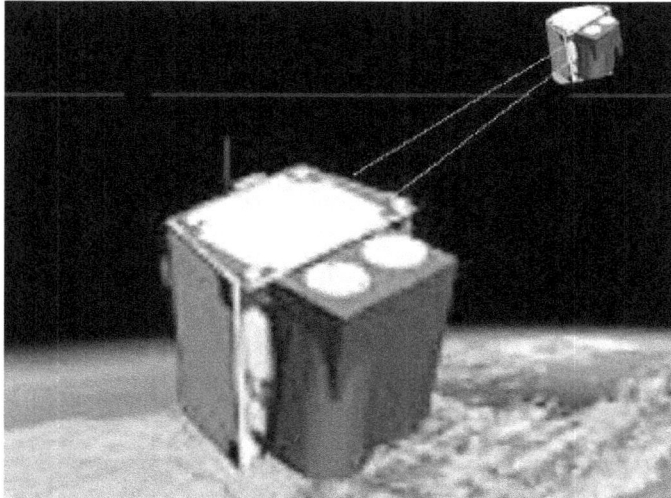

Figure 43. Liquid Droplet Thruster Concept for Formation Flying[35]

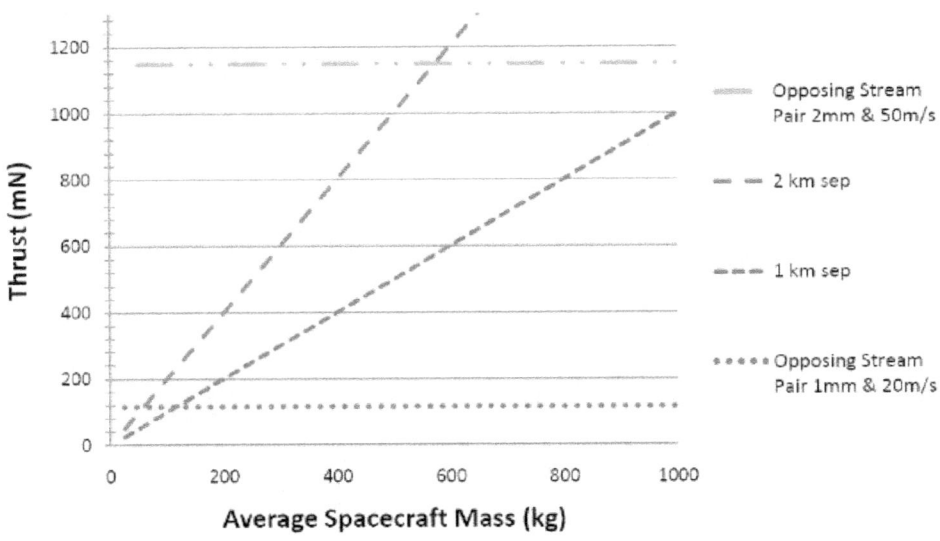

Figure 44. Thrust Required to Maintain Satellite Separation[35]

The liquid droplet thruster research investigated various aspects of generating, transmitting, and collecting such a droplet stream to determine the viability of the concept. The transmission of such droplet streams through space under environmental forces such as atmospheric drag and electrostatic interactions due to droplet charging was the main concern and was addressed using simulation. The droplet charging level was modeled using NASCAP. Many of the required droplet formation and collection technologies are similar to that developed under the NASA/USAF Liquid Droplet Radiator (LDR) program[36]. In LEO the primary concern for droplet transfer is drag while in polar earth orbit the primary concern is droplet charging and electrostatic effects. In a 300 km LEO orbit, drag can cause a droplet drift of 10 cm or more which must be compensated for. The net charge acquired by a droplet ranged from –26V to +21V in an 800 km polar orbit. Droplets were also analyzed for their potential to break apart due to the charge level and it appears that droplet breakup due to electrostatic charging is unlikely. Uncharged transit analysis using a Monte Carlo approach for the initial scatter indicated that the scatter for a typical system at a separation of 100 m is acceptable (within a 30 cm diameter circle).

Many significant perturbations were identified during the research, but no show-stopping effects were uncovered for this concept. Droplet streams are capable of providing several Newtons of thrust and are capable of maintaining a satellite ($m_s \sim 1,000$s kg) separation of more than a kilometer. As a distinct advantage, a conceptual droplet thruster system would weigh about one fifth as much as a comparable ion engine and consume about 1000 times less power. The relatively low required operating power of droplet stream propulsion effectively makes it an enabling technology for side-by-side tandem formation satellites. In addition, a droplet stream propulsion system contains most of the components needed in a LDR. Many components such as collectors and pumps were developed and tested as part of the LDR program and are well suited to droplet stream propulsion without modification.

The amount of off-course drift that droplets experience due to drag is affected by droplet diameter and speed that can be chosen to minimize off-course drift of droplets. It is very important that the droplet generation process yield similar sized droplets. During the research effort, microsolenoid valves were found to produce droplets of sufficient uniformity, size, and speed. Charging in Polar Earth Orbit (PEO) was found to be somewhat benign even during increased geomagnetic activity. The level of charging was found to be acceptable from the standpoint of unwanted droplet breakup due to the Coloumb repulsion. The major influence of both drag and charging on liquid droplets was found to be in the design of the collecting system on the opposing spacecraft. Relatively large collecting diameters and reasonable pointing control are required for this concept to be viable. Additional details are contained in the reference by Joslyn[35].

4.9 High Energy Advanced Thermal Storage (HEATS)[37]

In the High Performance Microsatellite Review (section 4.1), it was shown that Solar Thermal Propulsion (STP) offered distinct advantages as a propulsion system, providing a unique combination of high *Isp* (and therefore high ΔV capability) and relatively high thrust (for a fast response time). This technology showed considerable potential for enhancing the capabilities and enabling the use of microsatellites in a variety of missions. However, no STP mission has flown; this is due to a variety of reasons. One key reason that STP is not favored in satellite system designs is that it requires its own dedicated power system (i.e., the solar collection and concentration system), while other propulsion systems can utilize the solar panels and batteries already required on board. Additionally, a basic STP unit requires sunlight in order to heat the propellant and provide maximum thrust, effectively limiting propulsion maneuvers to times when the satellite is not in eclipse. In past systems that proposed thermal energy storage, the sensible heat of a warming and cooling block of solid material was suggested, resulting in significant temperature fluctuations that complicated system design and performance analysis. In order to mitigate these issues, the HEATS project seeks to combine a robust thermal energy storage solution, utilizing a phase-change material for relatively constant-temperature operation, in combination with a high performance thermal-to-electric conversion system. Such a system, illustrated in Figure 45, would enhance the flexibility and performance of the STP system, while simultaneously providing power output for other satellite systems and eliminating the need for the traditional power system.

The HEATS project requires a multi-faceted analysis, including selecting the phase change thermal storage material (PC-TSM), designing a container that can survive in contact with the thermal storage material (TSM) and not significantly react with it chemically, developing an advanced insulation system to retain the thermal energy, selecting and optimizing the electric conversion system, and developing the means of thermal output for propulsion. There is also constant systems analysis, mission analysis, and thermal analysis ongoing to analyze various system components and relative performance. This work has been ongoing, in both experimental form and in the form of theoretical and numerical analysis, in house and in cooperation with researchers at other AFRL branches, at the USC, at the UCCS, and at the Jet Propulsion Laboratory (JPL).

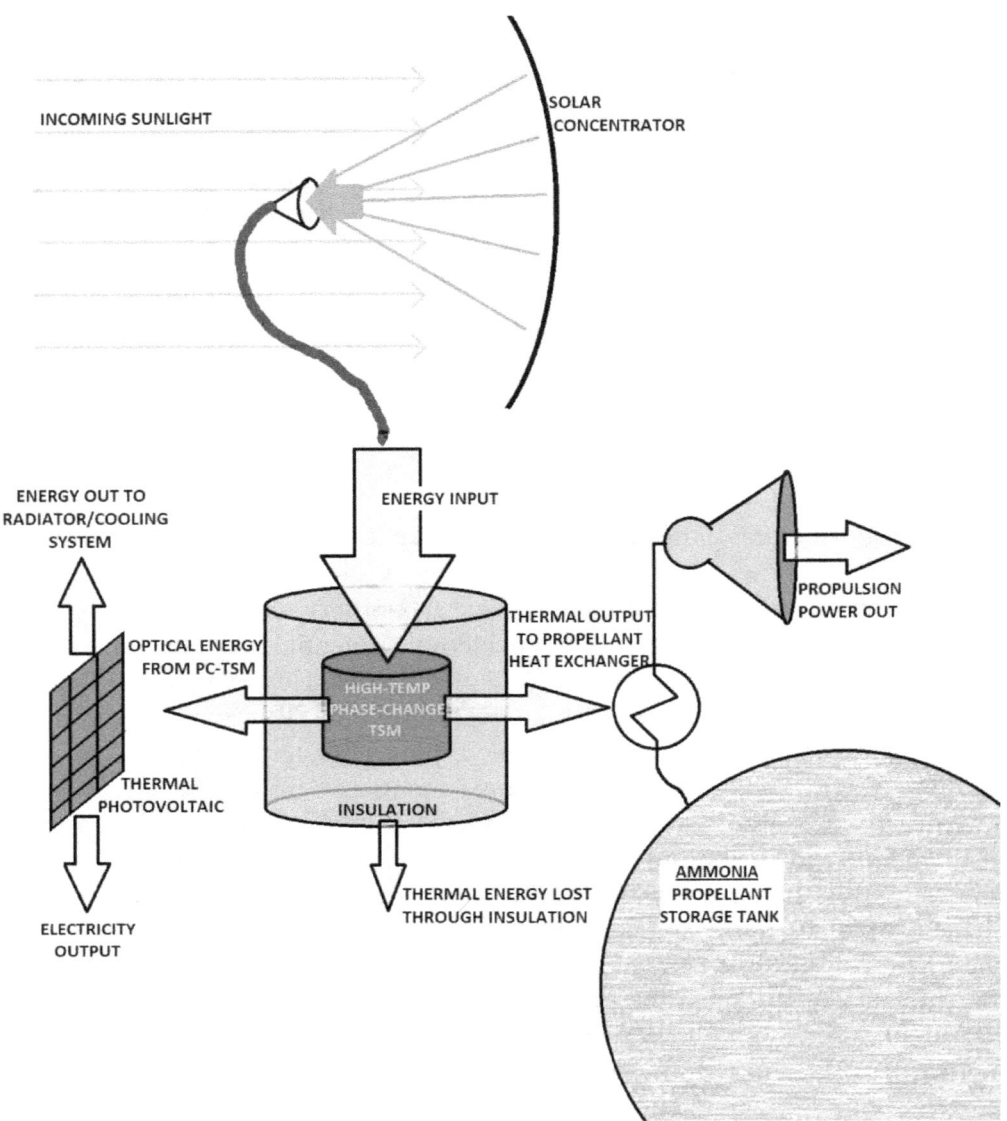

Figure 45. Solar Thermal Propulsion System, Augmented With Thermal Storage and Thermal-to-Electric Conversion.

In order to enhance system performance and robustness, analysis has shown that elemental phase change materials with high melting temperatures and high latent heats of fusion may be the optimal choices for storing thermal energy. Silicon (1687°K melting temperature and heat of fusion equal to 1785 kJ/kg) has been selected as a relatively near-term, moderate performance material, while Boron (~2500°K melting temperature and 4800 kJ/kg heat of fusion) has been selected as a farther-term high-performance material. A literature search and initial experimental analysis shows that graphite and boron nitride may be good crucible materials for containing the PC-TSM.

Thermal analysis has shown that state-of-the-art Carbon-Bonded-Carbon-Fiber (CBCF) comes close to the thermal resistance required to effectively insulate the system, and advanced Multi-

Layer vacuum Insulation (MLI) will certainly be adequate, although complex to work with. Collaboration with JPL indicates that refractory aerogels, currently under development, may be the ideal insulator. Thermal-photovoltaics have been selected for converting radiated thermal energy from the TSM to electricity. Ammonia has been selected as the propellant. Mission-based analysis shows that, for a high-ΔV microsatellite, the augmented STP system proposed here poses distinct advantages, allowing a notably higher ΔV compared to competing technologies operating within similar mass budgets. Future work will continue on this project, seeking to further understand materials interaction (especially under longer term testing) between the container material and the phase-change material (PCM), to scale up the solar furnace to allow higher temperature testing, to further develop advanced insulation materials, and to continue the various analyses to verify system benefits and performance capabilities.

4.10 Laser Fusion Concept[37]

Thermal neutrons with energies on the order of 0.025 eV are of considerable interest to the Department of Defense and for commercial applications. Unlike high-energy photons, neutrons easily penetrate high density targets, but get effectively absorbed by low density materials like paraffin, nylon, or explosives. This makes them attractive complements to X-rays for radiographic applications, e.g. for the detection or inspection of explosives inside steel casings. The key challenge, however, is: *the development of a compact thermal neutron generator with a large enough flux (e.g., 10^8 nuetrons/cm^2) to be useful for radiographic applications.*

The limited availability of radio-isotopes, combined with the relatively short half-life, safety constraints and regulatory requirements make them unattractive for wide-spread use as neutron sources. An alternative design exploits the Deuterium-Tritium (D-T) fusion reaction, which generates Alpha particles and fast neutrons. In these sources, Deuterium ions are accelerated to about 130 keV and hit a Tritium target. In order to be useful for radiographic applications, the 14.1 MeV neutrons must be thermalized in an external moderator, which reduces the functional neutron flux by two orders of magnitude.

The acceleration of Deuterium ions is usually accomplished in a diode configuration. Recently, considerable success has been achieved in the acceleration of ions via laser-matter interaction. In this project, therefore, the use of laser-accelerated ions to produce neutrons via nuclear fusion in an adequately designed target was investigated. The plasma simulation code VORPAL was enhanced with a model for fusion reactions and the generation of neutrons in shaped D-T targets was investigated.

It was found that neutron fluxes large enough for radiographic applications can be generated by utilizing moderate (10^{17}–10^{18} W/cm^2) laser pulse intensities and an appropriately shaped D-T mix target. However, it was also noted that a large portion of the target will not participate in the fusion reaction, and is therefore wasted. Additional modeling showed that, with changes to the target, the focusing of ablated material can lead to a high-density ion population suitable as a driver for Inertial Confinement Fusion.

Full details of this project can be found in the report: AFRL-RZ-ED-TR-2007-0065.[37]

5.0 SUMMARY

The Advanced Concepts effort worked to "enable future Air Force missions through the discovery and demonstration of emerging revolutionary technology." Reviews of technological fields were conducted for launch, near-space, and in-space propulsion. The reviews yielded a set of key technological challenges that were of particular importance for the effort:

1. *Reduce the cost of access to space for the entire launch spectrum by one order of magnitude at current launch rates.*
2. *Demonstrate a new propulsion mechanism that allows sustained low-speed flight over the altitude range from 40 km to 100 km.*
3. *Demonstrate a practical propulsion system that can operate efficiently ($\eta > 50\%$) at a specific impulse of 1000s.*
4. *Demonstrate a satellite propulsion system that achieves high thrust efficiency over the entire range of specific impulses from 500s to 5000s.*
5. *Individually demonstrate the components that could enable ambient gas to be utilized as propellant for low altitude orbiting systems.*
6. *Demonstrate a microsatellite satellite propulsion system that could provide a total mission ΔV of 1.0 km/s with a response time of days.*
7. *Demonstrate a propulsion system that can allow satellite formation flying at significantly higher effective specific impulse ($Isp_{eff} > 100,000s$) and avoid contamination of other satellites in the constellation.*

The small scope of the Advanced Concepts effort allowed only small, targeted research efforts towards addressing the key challenges, but significant progress was made towards addressing several of them. Table. 8 lists a summary of the technologies that were investigated and their current status. Advanced Concepts research is, by nature, highly unpredictable so none of the concepts are "thrown out", but concepts that appear to have fundamental challenges are "paused" awaiting major improvements. Both the microwave augmentation of solid rocket motors and the EHD thruster fit in this category. Early performance estimates predicted that the microwave augmentation concept would yield only modest increases in performance, but would require a very costly ground station. Efficiently beaming microwaves through the exhaust and to the expanding portion of the nozzle also appears difficult. This project awaits significant reductions ground station costs to make additional research into the efficient absorption of the microwave energy worthwhile. Early modeling efforts concluded that the EHD propulsion effort would require significant reductions in the ion mobility to increase the coupling between the ion and neutral flows to make the device practical for near-space operations. Early reviews of potential methods for achieving the reduction did yield some improvements, but not at the required scale.

Several of the individual concepts were particularly strong candidates towards meeting the key challenges: radiometric force propulsion, liquid droplet thruster, and HEATS. The research conducted in the radiometric force propulsion project assembled, for the first time, a general understanding of the radiometric force mechanism that would allow the calculation of the force for somewhat arbitrary vane arrangements. This new understanding was used to estimate the performance of radiometric thrust plates in near-space and the results look promising. The project continues with the proof-of-concept of a scaled, but relevant, thrust plate in vacuum expected within a year (currently 2012). The liquid droplet thruster answered all initial concerns about the concept using both modeling and experimental techniques. Early designs showed that the concept could achieve the required goals. The next step in the research is simply to demonstrate the full system in a vacuum. The HEATS project has just started, but early systems estimates look promising. Background literature searches have also indicated that there is strong potential for the concept to be able to achieve the required performance levels. One concept, FRC, has been successfully transitioned and is in the process of being developed into a complete thruster package.

Table 8. Status of Researched Advanced Concepts Technologies

Technology	Challenge	Status	
		Research	Requirements
μW Augmentation of SRM	1	- Initial Analysis Complete. - Minor Improvements Expected. - Project Paused.	- Cheap μwave ground station. - Efficient coupling through exhaust.
EHD Propulsion	2	- Initial Analysis Complete. - Significant Advancements Req. - Project Paused.	- Significant Reduction in Ion Mobility.
Radiometric Force	2	- Basic Research Nearly Complete. - Promising Systems Calculations. - Performing Perforated Vane Experiments.	- Proof-of-Concept Demonstration of Thrust Plate.
mHTX	3	- Initial Research Complete. - Promising Early Results. - Project Paused.	- Increase Efficiency. - Packaged Device.
Capillary Discharge	3	- Completed Early Research. - Demonstrated Ignition Technique. - Nearly Achieved Required η_{th}. - Project Paused.	- Continue Increasing η_{th}. - Demonstrate Propellant Feeding. - Full Device Demonstration.
MLP	3,7	- Early Research Started. - Showed Low Speed Operation. - Project Paused.	- Optimize Coupling to Macron. - Incrementally Increase Macron Velocity.
FRC	4	- Early Research Complete. - Technology Very Promising. - Project Transitioned.	- N/A
Pyroelectric Thruster	*Unknown*	- Preliminary Research Complete. - Work Halted Before Demonstration.	- Initial Electrospray Demonstration. - Systems Level Estimates.
Liquid Droplet Thruster	7	- Initial Research Complete. - Basic Concept Shown Viable. - Project Paused.	- Ground Demonstration of Closed System (Including "Pitch" and "Catch")
HEATS	6	- Project Initiated. - Early Systems Estimates Show Clear Advantage. - Project Ongoing.	- Demonstrate Long-Term Containment of Phase Change Material. - Complete Design with Full Concept of Operations.

www.ingramcontent.com/pod-product-compliance
Lightning Source LLC
Chambersburg PA
CBHW080543190526
45169CB00007B/2621